I Went Aviatin' to China

My Father's WWII Letters to My Mother

Richard Eric Pound

High Tide Publications, Inc.
It's never too late to write

First Edition

ISBN: 978-1-962935-06-7

Published by High Tide Publications, Inc.
www.Hightidepublications.com
Deltaville, Virginia

Edited by Cindy L. Freeman (cindylfreeman.com)

Book design by Firebellied Frog Graphic Design (firebelliedfrog.com)

This book is dedicated to my father,
whose letters tell the story,
and to my mother,
who saved the letters
and passed them on to me.

Table of Contents

A Note to Readers:

This story is built on letters from World War II. It conveys the words, language, slang, attitudes, and expressions of that period. Some of those may be uncomfortable or even offensive to our modern ears. I haven't changed anything. I will not substitute my words for my father's words eighty years after the fact. He has the right to be heard in his own voice.

> "The past is a foreign country; they do things differently there."
> — L.P. Hartley

Publisher's Note:

The font used in this book for the letters written by Joe Thomas Pound is American Typewriter. We use this font because it mirrors the type from letters and documents written during World War II. The author, out of respect for the original text, uses square brackets to comment on his father's words. The text is encapsulated in square brackets [ ]. The parenthesis text ( ) in the letters is written by the author's father and is part of the original letters.

TERRY AND THE PIRATES
by MILTON CANIFF
LET'S TAKE A WALK, TERRY...
YES, SIR, COLONEL CORKIN!
I'M GOING TO MAKE A SPEECH—AND IT'LL BE THE LAST ONE OF ITS KIND IN CAPTIVITY—SO DON'T GET A SHORT CIRCUIT BETWEEN THE EARS...
NO, SIR
WELL, YOU MADE IT...YOU'RE A FLIGHT OFFICER IN THE AIR FORCES OF THE ARMY OF THE UNITED STATES...THOSE WINGS ARE LIKE A NEON LIGHT ON YOUR CHEST...I'M NOT GOING TO WAVE THE FLAG AT YOU—BUT SOME THINGS YOU MUST NEVER FORGET...
...EVERY COUNTRY HAS HAD A HAND IN THE DEVELOPMENT OF THE AIRPLANE—BUT AFTER ALL, THE WRIGHT BROTHERS WERE A COUPLE OF DAYTON, OHIO, BOYS—AND KITTY HAWK IS STRICTLY IN NORTH CAROLINA... THE HALLMARK OF THE UNITED STATES IS ON EVERY AIRCRAFT...
...SO YOU FIND YOURSELF IN A POSITION TO DEFEND THE COUNTRY THAT GAVE YOU THE WEAPON WITH WHICH TO DO IT... BUT IT WASN'T JUST YOU WHO EARNED THOSE WINGS... A GHOSTLY ECHELON OF GOOD GUYS FLEW THEIR HEARTS OUT IN OLD KITES TO GIVE YOU THE KNOW-HOW...
...AND SOME SMART SLIDE RULE JOKERS SWEAT IT OUT OVER DRAWING BOARDS TO GIVE YOU A MACHINE THAT WILL KEEP YOU UP THERE SHOOTING... I RECOMMENDED YOU FOR FIGHTER AIRCRAFT AND I WANT YOU TO BE COCKY AND SMART AND PROUD OF BEING A BUZZ-BOY...
...BUT DON'T FORGET THAT EVERY BULLET YOU SHOOT, EVERY GALLON OF GAS AND OIL YOU BURN WAS BROUGHT HERE BY TRANSPORT PILOTS WHO FLEW IT IN OVER THE WORST TERRAIN IN THE WORLD! YOU MAY GET THE GLORY—BUT THEY PUT THE LIFT IN YOUR BALLOON!...
...AND DON'T LET ME EVER CATCH YOU BEING HIGH-BICYCLE WITH THE ENLISTED MEN IN YOUR GROUND CREW! WITHOUT THEM YOU'D NEVER GET TEN FEET OFF THE GROUND! EVERY GREASE MONKEY IN THAT GANG IS RIGHT BESIDE YOU IN THE COCKPIT—AND THEIR HANDS ARE ON THAT STICK, JUST THE SAME AS YOURS...
...YOU'LL GET ANGRY AS THE DEVIL AT THE ARMY AND ITS SO-CALLED RED TAPE...BUT BE PATIENT WITH IT... SOMEHOW THE OLD EAGLE HAS MANAGED TO END UP IN POSSESSION OF THE BALL IN EVERY WAR SINCE 1776—SO JUST HUMOR IT ALONG...
14th AIR FORC
U S
OKAY, SPORT, END OF SPEECH... WHEN YOU GET UP IN THAT "WILD BLUE YONDER," THE SONG TALKS ABOUT—REMEMBER, THERE ARE A LOT OF GOOD GUYS MISSING FROM MESS TABLES IN THE SOUTH PACIFIC, ALASKA, AFRICA, BRITAIN, ASIA AND BACK HOME WHO ARE SORTA COUNTING ON YOU TO TAKE IT FROM HERE! GOOD NIGHT, KID!
This way to TOKIO!
Next stop U.S.A.

A Comic Strip by the Door

For several decades a photographic reproduction of a famous *Terry and the Pirates* comic strip episode has hung by the front door of our home. The episode was dear to my father. Here are his words:

Note for Heirs and Assigns:

This is a photographic reproduction of what is perhaps the most noted single comic strip episode in this century. Published in December 1943, it received much comment in the press and the Congress at the time and was read into the Congressional Record. I have seen reference to the strip in print many times in the intervening years. Most graying WWII aviators are familiar with the strip.

At the time of publication, I was at the San Antonio Aviation Cadet Center (now Lackland AFB) awaiting orders to Primary Flying School. I sent the strip to my mother who stored it in a dark place until the war ended. For 35 years my wife (and dearest friend) has kept it in a light-free environment. The original paper strip is stored in this frame behind the copy. It is unfaded and unyellowed. The color negatives are attached.

The Colonel Flip Corkin in the strip (called Colonel Corky by his men) is closely based on a real-life legendary USAAF Colonel named Phillip Cochran who was widely respected in the China/Burma/India theater of war at the time.

Colonel Corky made special arrangements with the War Dept. for Terry to receive Aviation Cadet training in China (Kunming) under his supervision. This is the big day that Terry received his wings. Please treasure it as I have.

Love Paw
Joe T. Pound
24 October 1981

Addendum:

About 14 months after this strip was published I was in India flying cargo aircraft (C-46) across the Hump to China. Seventy-two trips of which about 65 were into Kunming.

When the war ended I had 1300 flying hours of which 400 were in weather.

Dad passed away three years later 8 October 1984. After he died, Mom gave me the reproduction, along with the original, the negatives, the frame, and Dad's note. She said he wanted me to have it. I treasure it as my father did.

Foreword

Shortly before she passed away in 2016 at the age of 90, my mother gave me some packets of letters she had saved. Among them were the letters my father had written to her during World War II. She said she would like me to "see what you can do with these." This book represents my attempt to do that.

While I was growing up, my father was a Senior Field Examiner for the Indiana State Board of Accounts — in other words, an auditor. He was also a pilot and a senior officer in the US Air Force Reserve. The former required him to be away from home regularly; the latter required two weekends per month. The upshot was that Dad and I spent little time together during my boyhood.

Then I was off to college and graduate school, got married, had children of my own, and embarked on a career that took me overseas. Dad died young, at 59. He was a fine and admirable man, but it's fair to say I never knew him as well as I should have or would have liked. Our children and grandchildren (and nieces, nephews and generations to follow) are the audience for this collection, but in the first instance it was my own experience of discovery. The letters comprise both a war narrative, one of many that are similar, and a love story, singular and unique.

All her life my mother kept the small local newspaper clipping that reported her marriage to Dad. We found it in her wallet after she passed away. For such an admirer of Dorothy Parker and advocate of women's independence and liberation long before that was even a term, it has been remarkable to learn how fully she devoted herself to Dad, their marriage, and our family. That is love. Dad left India in November 1945, and he and Mom married on 26 January 1946. There are more chapters to their story after that, but those are for later.

My father, Joe Thomas Pound, graduated from Sullivan High School on a Friday in May 1943. The following Monday he was on a train to begin basic training in the U.S. Army Air Forces (USAAF). He told me once that his mother, Nellie, ran beside the train as it pulled out, waving goodbye. He said his thought at the time was: "Please stop, Mother." He was able to enter USAAF training so soon after graduation because he had already taken all his physical examinations and mental aptitude tests while he was in high school.

My dad between two high school friends

The U.S. had already been in World War II for a year and a half, and Dad —like his classmates—knew what was in store when he graduated. He had always wanted to fly, and now he would get his chance. President Roosevelt had committed the country to train 50,000 pilots. Joe Thomas Pound was to be one of them. He had lots of company; from 1 July 1939 to 31 August 1945 the U.S. Army Air Corps and its successor USAAF trained 193,440 pilots.

Dad and life-long friend Bill Able

Dad died on 8 October 1984. My mother, Mary Jane Griffith Pound, outlived him by more than thirty years. Before she passed away, she gave me the letters between her and Dad from during the war, before they were married. In fact, these are all letters from him to her, which she saved. I don't know whether he saved any of hers, but none came down to me. I'm told this is not all that unusual since men in war often focus on fighting and surviving more than preserving artifacts for a future they may never see. Regardless, the record is heavily about him, although some of her experiences may be inferred from his references to her correspondence.

Dad's high school graduation picture

To answer my questions after Dad died, Mom wrote down her recollections on 17 October 1986. I have quoted from her 1986 narrative where appropriate.

Mom and Dad went to the same high school but were friends rather than boyfriend and girlfriend. Mom graduated in 1944. I always wondered whether they had an eye on each other early on, and there is some evidence for that, but Dad's letters refer to only a single date. They married soon after Dad returned from the war, so it seems the letters comprised their courtship. Mom was offered a scholarship to Smith College but after graduation chose instead to go to Dayton, Ohio to support the war effort by working for the Veterans' Administration (VA). It was there that she received from Dad an envelope postmarked 25 July 1944.

The envelope contained a formal announcement of the graduation at "Blackland Army Airfield of Waco, Texas," on "Friday morning, August fourth, nineteen hundred and forty-four, at nine o'clock, Post Theatre of USAAF Class 44-G."

Enclosed with the announcement was a calling card for "Joe T. Pound, Flight Officer, Air Corps, Army of the United States." Flight Officer was a rank specially created with Roosevelt's new pilots in mind; it was more junior even than 2nd Lieutenant and I suspect was meant to allow a bit more time to evaluate the herd of new flyboys before trusting them with expensive aircraft in combat theaters.

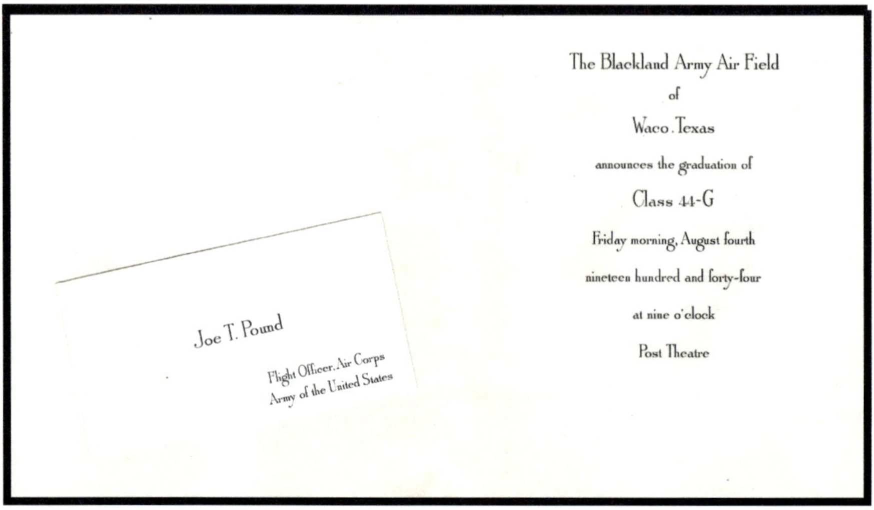
Joe T. Pound

Flight Officer, Air Corps
Army of the United States

The Blackland Army Air Field
of
Waco, Texas
announces the graduation of
Class 44-G
Friday morning, August fourth
nineteen hundred and forty-four
at nine o'clock
Post Theatre

So, where had Dad been during the roughly fourteen months between his entry into the USAAF in May 1943 and his graduation in August 1944? In 1986 Mom recalled the following: *I can give you a general idea of what happened. He went from Ft. Benjamin Harrison, Indianapolis, to Keesler Field, Mississippi, for basic training. He arrived there early summer of 1943. Summer in Mississippi without air conditioning is not a pretty sight. The first night there, he was assigned to KP. Mountains of potatoes to peel — probably for the first time in his life. The mess hall had to be scrubbed down. The one plus for the crew was the fact that there was a limitless amount of ice and lemonade available to them. He partook freely and it gave him a raging case of hives. Not only was he lonely, homesick, hot, sweaty and tired — he itched all over. He survived basic in the heat of Mississippi and the alternating mud and dust and proceeded from there to Western Reserve [University], Cleveland, Ohio.*

Cleveland in those days was a thriving industrial center. Western Reserve was Joe's first contact with a university setting. He liked it. He could do the work. The credits he earned there transferred to IU [Indiana University] when he started college in 1947. This was an academic oasis in the rush to war. A time of crisply creased uniforms and proudly shined shoes.

However, when they arrived there on the train from Keesler it was a different story. A word about wartime trains: most of them were coal burners and they were not air conditioned. The passenger had the choice of opening the windows and being blanketed with the soot and cinders which already covered the train and seats or closing the windows and suffocating. The trains were always crowded. Lucky people got seats. Others sat on the suitcases or stood. Sometimes a soldier or sailor traveling a long way simply lay down in the aisle and slept among the soot, gum wrappers, and other debris. This is why when Joe stepped off the train in Cleveland he was rumpled, unshaven, and dirty.

They were met at the station by a contingent of cadets from the university. The cadets, of course, were perfectly groomed. One of them made it his personal mission to put Joe in a brace and inform him of his deficiencies in appearance. Joe could never remember exactly what this cadet looked like, but he had an excellent opportunity to study his name tag. It was CATLIN. Years later, during the early '70s, he worked with a Col. Ben Catlin. He did not make the connection for some time, but eventually he and Ben concluded that Col. Catlin and the overzealous Cadet Catlin were one and the same.

Western Reserve University became Case Western Reserve University in 1967 when it federated with Case Institute of Technology. So the educational experience at what is now known as Case Western Reserve University comprised a big chunk of Dad's time in 1943 and early 1944. Dad once told me that even with the exigency of war the USAAF took time to provide cadets with a highly concentrated and accelerated dose of university education. They were to be officers and gentlemen, after all. Now I know where that happened. Dad's USAAF Separation Qualification Record (SQR) of 14 December 1946 reads as follows: "College Training Detachment, Western Reserve Univ., Cleveland, Ohio, Air Crew Training (Physics, Math, English, History, Literature). Five Mo. course completed in two months. July and August 1943."

Dad's address book includes two Cleveland entries. The first is for Robert C. Ensminger, 9205 Rosewood Ave., Cleveland, Ohio. The second is an entry for Miss Jean Kodrick of 2007 W. 20th, Cleveland. Telephone is PROSPECT-7654. Her entry is notable for including the comment "WOW!"

The SQR next says Dad went to "Pilot Preflight School, San Antonio,

Texas (Physics, Math, Code, etc.) 9-wk course completed in December 1943." This was at what was then known as the San Antonio Aviation Cadet Center. That is all the detail I have because a fire at the National Personnel Records Center on 12 July 1973 destroyed Dad's records and many others.

Joe T Pound AAFTD Fort Stockton, Texas 44-G Primary Training 1944

What I do know is that Dad was subsequently part of training class 44-G, the seventh training class of 1944. They started learning to fly under the watchful eye of the Army Air Forces Flight Training Division (AAFFTD) at Fort Stockton, Texas. The actual flying for primary training took place at Gibbs Field, about one mile north of Fort Stockton. The class book includes a photo of Dad, dashing in his fleece-lined leather jacket, helmet, goggles, and white scarf. His call sign was "Flash," a carry-over from a part he played as a news photographer, in his high school senior class play. I was told that many years later, when the old high school building was demolished, a brick in the backstage area of the auditorium had on it the graffito "Flash 1943."

Then it was on to Curtis Field near San Angelo, Texas for basic flight training. The USAAF started with "primary" training to see whether the cadets could get off the ground and back down in one piece. Next, he attended "basic" training for increasingly advanced aircraft and flying tasks. Then came "advanced" training at Blackland and, finally, that 4 August graduation.

At some point while at Fort Stockton, Dad wrote a letter that highlighted his dislike of Texas. From the content, I could tell he wrote it after he had been in Texas for a while and after Mom had gone to Dayton to work for the VA. Also, it seemed this was not the first letter he wrote to her.

Same Time, Same Damn Place

Hello You – Just to show you that (1) I've your best interests at heart and (2) am interested in keeping your morale (whatever that is) high with a full mail box and (3) always pay my correspondence debts (This is a sort of a Three Point Plan) I'm answering Letter No. II which was by the way very interesting. I especially liked the "meooww," girls are funny people.

I don't know what we'll talk about because I've already told you everything I know and some I don't. I've been in Texas so long my main interests lie in dodging rattlesnakes and the large, economy size mosquitoes. All of which fills your mailbox but makes dull conversation.

You didn't tell me what you do or how you like Dayton etc. It would undoubtedly be wonderful to walk into a nice cool bar and order a frosty mixed drink rather than slip down two dark alleys turn right and hand Johnny $5 over the back fence (looking over your shoulder for constables etc.) for a pint of double potent varnish remover. The word hasn't gotten to Texas yet that Prohibition was repealed years ago. Most of the country is dry. They are still living in the days of Sam Houston and the Alamo. It was the last time Texas did anything that amounted to anything so they just keep talkin' about the "deah, Deah (!) old Alamo!" I saw it in San Antonio. Looks like a pile of rocks to me.

I love it here. The climate is so healthful – never more than 115 degrees!

This ink belongs to Pace. He calls it his purple potent passion fluid. Not Doug but Ken Pace. He is from Washington State and a pal of mine.

Ken Pace indeed appears in the Fort Stockton 44-G class book. Doug on the other hand was Dad's close high school friend, Doug Pace.

I use his ink and he borrows my clean towels. Big buddies.

I wish my boy Doug would write me. Have you heard from him recently? Have you heard from your brother [Wayne]?

I've gotta go now. Keep one eye on that nice boss and a good sharp pencil behind your ear. Goodnight Government Girl –.

Yours, Joe T.

Farm boy Joe - about 10 years old.
Ten years later he would be flying over "the Hump."

Getting Wings

The earliest letter from Dad that has come down to us with a date is 27 May 1944 to "Dear Mary." He opens by saying he would have written sooner but he was busy packing, unpacking, and getting settled "on a new post"—presumably Blackland. He also confessed that after finishing Basic he and three "pals" went to San Angelo ("not DRY") for what seems to have been a three-day bender ("a long-awaited celebration") at a local hotel. "We . . . kept two bellhops busy—one carrying up -er- liquid refreshment and the other carrying down empty bottles."

Then more seriously:

> This is it, Mary – the last lap, advanced. This is a twin engine school. Those wings are getting close now.

He then congratulates Mom on her high school graduation and notes with admiration that she was on the honor roll. Some of that page is illegible because of something spilled on it. Later he mentions a movie entitled *A Guy Named Joe* and notes it includes some good shots of a P-38. He says, "Maybe I can get a P-38 out of here but the chances are pretty slim. Most of the boys go to medium and heavy bombardment."

Next, he turns to his friend Doug. Dad hadn't had a letter from him for a while and he was concerned:

> Confidentially I'm worried about the guy. He's in a rough outfit and I have an idea he'll be in the first wave when the invasion starts. You know as well as I what the odds are. Why don't you write him a cheerful letter? He thinks a lot of you, Mary.

Finally, he writes about his schedule:

When we get started flying we're going to be pretty busy. Up at 5:00A.M., on the Flight Line all morning and ground school until 9:00 at nite. That leaves a whole hour every day for us to do as we please! Gee! Aren't the Army generous! So long, kid. Write. Joe T.

Doug Pace was indeed in a rough outfit—the U.S. Army Rangers. As expected, he was in the first wave on D-Day in Normandy. He was wounded twice but survived the war and lived to the ripe old age of 90, passing away in 2015.

The next letter is undated and unplaced, but I'm confident it was from advanced training at Blackland because of the reference to the AT-17. The Cessna AT-17 Bobcat or Cessna Crane was a twin-engine advanced trainer aircraft used during WWII to bridge the gap between single-engine trainers and larger multi-engine aircraft. The letter is addressed to "Dear Brat" and clearly picks up some ongoing kidding. By context, this one seems to fit after the 27 May letter but before the 7 July letter because there's no countdown of days until graduation as there is in the 7 July postscript. The stationery shows an AT-6 with a rabbit straddling the nose, asking "Hi Doc, What's cookin?"

Knowing you I thought you would like this stationery. It's the last I have. You owe me a letter, I think, but it isn't everyone who can write to a charming and intelligent young lady – (well charming anyway – ouch!) Now don't get the idea I like writing you, you stubborn brat. I just thought the wittle wabbit would amuse you.

I've been talking to a couple Brazilian cadets. They're in the Brazilian Army being trained here by the United States. We have about twelve in our class getting the same stuff we are. Two of them live in my barracks. Their native language is Portuguese but most of them speak better English than I. Which isn't, I believe, too great an accomplishment. But on the whole they're a bunch of pretty intelligent joes. They like Frank Sinatra too.

Tomorrow's Sunday but we're going to get up at 5:00 and fly in the morning. [Followed by symbols meant to represent cursing.]

We have AT-17s here. They fly about as well as a shoe box but I won't be flying 'em forever. I'll send you a picture of one when I get it developed.

By the way, who's the love interest as of this week? Who is the latest victim of the wicked vampress?

I mentioned before that it's Saturday nite and time for the weekly shave and shower.

No cracks or I'll break your pretty little neck when I see you.

Adios Señorita, Joe T.

The next letter is from Blackland Army Airfield, Waco, Texas, 7 July 1944 to "Dear Janie." Mom had apparently sent him a photograph of herself.

Ya know what? I think ya need a haircut, gal. And by the way – can you still raise that right eyebrow in that go-blow-soap-bubbles-you-big-lug manner? And as I remember you were always picking on me. You just took advantage of me because I was young, that's all. And I remember a time when you deliberately tore up a box of starch and scattered it all over Herman's nice clean bus not to mention a dozen or so of the smaller fry of the McMahan species. [Their school bus driver was Herman.]

Aren'tcha ashamed? And I wonder if that bridge still goes over the water?

Yeah, you still got that gleam in your eye. Thas (sic) all I wanted to know. And I like the picture whether you do or not. My bunk mate likes it too. I'm going to have to do something with that guy.

Had a letter from Doug this week. I think we should write him don't you? He told me he had been hit in the finger by a sniper's bullet but he didn't say it was serious. Anyway he can still write.

Didya ever hear from your brother? Incidentally, Doug's address is still the same.

Gee I'm tired. I was up 3 hours with my instructor this afternoon. I don't like to fly with my instructor – he makes me work.

We flew last nite, too. I didn't get in bed until 3:00 this morning. Tomorrow nite (being Saturday, of course) we have to fly up to Abilene and back in the dark. They would pick a night like Saturday for that. It just wouldn't be right to go on Tuesday. I can't see how they expect the liquor stores downtown to escape bankruptcy if they persist in keeping me in weekends.

Which reminds me of something else. Out on the flight line they have a little racket called the Star System. There's a big bulletin board with all the cadets' names on it. Every time one of the boys pulls a knucklehead stunt while flying (such as bouncing a landing, etc.) his instructor puts a nice little star up by his name. At the end of the course we have to pay 25 cents for every star we have. And the money goes to buy -er- food and stuff for the graduation party we're going to have with the instructors. You know me. Honest, Mary, I ain't any dumber than anyone else – I just get caught oftener that's all. But it looks as [if] I'm going to be the leading contributor to the liquor fund. Hell, they should have made me Chairman of the Finance Committee to begin with. Now how'm I goin' to drink all my $9.75 worth of beer in one evening?

I gotta go see a man about a bed now. I'll be seein' you. Write.

J.T.

Postscript: 27 more days!

The next letter is undated. Where the date would be is the notation "21 more days." Given the postscript above, I'm pretty sure that's twenty-one more days until the end of advanced training and Dad earning his wings.

Dear Mary

You know it's awfully hot to be writing letters. And I'm going to have to fly tonite to 2:00AM so I could be sleeping. But if I don't write you, you wouldn't answer. So see all the sacrifices I make for one little letter from you?

It's a lovely afternoon in Texas. So nice and warm. Only about 115 degrees! I've been spending a quiet afternoon taking 17 cold showers. Dull isn't it? and this guy [Glenn] Miller has the nerve to play "Moonlight Serenade."

Woman, neither fear of God nor that gleam in your eye could frighten me enough to persuade me to remain in Texas. I think we should get a service ribbon for duty in Texas.

Sorry you want me to bring my honorable bunkmate home with me. No can do. He's taking the first plane to Virginia to get married. Poor misguided lad!

Sorry, too, but couldn't take you to the graduation party. It's strictly stag. The major always wears his oldest uniform – he invariably gets thrown in the river. But if you'll come to Texas perhaps I could be persuaded to skip the graduation party.

My roommate just made a very, very intelligent crack. He was lying all rumpled on his bunk and pulled his head out from under the pillow, opened one eye and remarked – "Warm isn't it?" It took a genius to figure that one out.

Then, "4 days later, 17 more days, to go" he continued...

Awright Awright! So I ain't neat!

On bended knee I offer my most humble apology and beg of you, fair damsel, to view the facts before placing the curse of the Crimson Sash upon my downtrodden brow. [The origin of the "curse of the Crimson Sash" is lost to history.]

I think I'll have a portable shower and a collapsible bunk installed in my airplane – I've been living in it for four days.

One oh-so-black nite we flew down to Austin and back. About 25 planes went down. We were checked out here at 5 minute intervals so we couldn't follow each other. 24 of us got back. One guy had engine failure and scattered an AT-17 over 40 acres of Texas countryside. Both boys in plane walked away from it.

And then Saturday I went on a low altitude cross-country to Jacksonville (Tex.) with my bunkmate. We flew at 200 ft. and had to dodge thundershowers all the way.

Went swimming yesterday (Sun.) afternoon. Bunkmate remarked casually that "the water was rawthuh wet."

And out in Virginia he doesn't live in a house – he lives in a "haiuse." Back home a "howse" is something to ride.

When will these provincial Eastern lads learn to speak English?

Later today – Hi You – Just got back from lunch and have to leave for the Flight Line any minute now so I'd better wind this up and get it in the mail or it will be on cold storage four more days. Write Brat – Luf, Joe T.

This was the first use of the "Luf" closing but would not be the last.

And then on 4 August, Dad became "Joe T. Pound, Flight Officer, Air Corps, Army of the United States."

He was 19 years old.

Joe T Pound - Curtis Field Class 44-G Basic Training

Wearing the Wings

Dad's next letter (on USAAF stationery) is dated 13 September 1944, location unspecified. Dad's SQR put him at Goodfellow Field, San Angelo, Texas, in August and September 1944. Anyway, there's a discernible shift in tone from previous letters.

> Dear Mary –
>
> Since this is the 13th I'm going to bed as soon as I finish this letter just in case. I don't trust Fate any more. Or women. (hmmm)
>
> Did you know Doug is back in the States? In case you hadn't heard he's at Wakeman General Hospital, Ward 18A, Camp Atterbury, Indiana. I think he's getting along O.K. He wrote my mother he's expecting a furlough before long. He had back and leg wounds and a piece of steel through his ankle. I wonder are there any cute nurses in Ward 18A? Not that I'd be interested or anything.
>
> I'll be danged. I just dug up your last letter and reread it and I'll be darned if you didn't tell me about Doug being back home. Oh well, he makes nice conversation anyway. You and he are two of my best pals, I hope. And whadda ya mean will I come around and say "hi" when this Big Mess is over? Watcha tryin' to do give me the brush off? I've been laying awake nites planning our future and you come along and say I can come around and say "hi" and then, I presume, beat it.
>
> If this persists I'll go out and shoot myself first thing in the morning.
>
> I'm having a helluva time reading this letter of yours. I

laid it on my windowsill and went happily in town and then the rains came – and the ink ran. My pal Ruley wants to go swimming at the pool in town this afternoon so I guess I'll humor him and go along. Right now he's sawing logs on my bunk. The noive of that guy cluttering up my unhappy little home. I have a swell little suite here at the BOQ [Bachelor Officers' Quarters]. It consists of a big, airy (9x9) room with 4 walls (floor too!) bed, table, a wardrobe, and your picture. I consider your picture an essential furnishing. If I get overseas before the war's over I'll carry it on every mission. One look from you used to keep me on my side of the fence – maybe you can impale all the gremlins on one of those icy glances.

According to Dad's address book, "pal Ruley" may have been Ben Ruley of El Monte, California.

Speaking of gremlins – did friend Tommy ever get home? (hmmm) I swear Mary I'm not at all jealous – much. [I do not know who "friend Tommy" was.]

I suppose I'll just have to go on carrying this torch. I do hope I don't get my fingers burned. Well – I'll check in for now. Love and Stuff, **Josef.**

This letter documents a romance that had already begun, at least on Dad's part.

The next letter is dated 27 October 1944 and was written on stationery headed: "Officers' Club, Goodfellow Field, San Angelo, Texas." This letter marks the end of one chapter and the beginning of another. I would not rule out the possible influence of alcohol.

Fair Lady –

Fate, in its realm of Destiny, has ofttimes been kind to me in my journey thru this that we call life. But more often than not the path has been littered with trials and disappointment. (Gee, ain't I eloquent?) Anyway, Mary, darnitall whydja hafta go back so so soon? At the first opportunity I escaped from my mother's little prison of loving care, dashed madly to a phone and – But Mary went back to Dayton yesterday. And so I say, my lovely one, long time no see and now longer still.

Once, in the dust covered files of history a very famous general remarked – "War is hell." And he ain't kiddin'.

And so I stroll into the old family wigwam quite unexpectedly — bowed by the weight of centuries (and three suitcases) and (after the preliminary kissing and crying) my mother grasps me firmly by the collar (with that fanatical gleam in her eye) and says with a fiendish chuckle — "Aahh, now we'll go out and visit all the relatives and show them how straight and strong and brown my boy has grown!" (oh my aching back!) And so, Mary, for 13 days I answered questions for Aunt Minnie and Uncle Rufus and Cousin Ignatz. I actually saw one fellow from the old gang. Namely Bill Able.

I guess it's no use trying to make up for a year in two weeks. Wars don't last always so maybe the sun will shine 'round our back door someday. Mary I didn't even call Gertrude and I'll be in the dog house from now on.

Despite much searching, I cannot identify Gertrude, who is mentioned often in these letters. She seems to have been well known to and friendly with both Mom and Dad. There is no Gertrude in the Sullivan High School classes of 1942, 1943 or 1944.

I'm not going to be here very long I'm sure. Fifteen of us have been alerted for two days to go out on secret orders. As hush-hush as the thing is it might be overseas. We don't even know yet where we're going. Anyway, Mary, please write and tell me what you're doing and all the local dirt etc. And by the way please don't become too interested in any particular one of those Air Force uniforms. Inside they're all just still big bums like me who used to be very troublesome nuisances when you were in high school.

But gee don't I wish I were back there throwing paper at you in the squeaky old bus — and whistling at Gertrude on the stairs — and tearing up boxes of starch — and flirting with Baldridge — and you, and exchanging tales of woe about our various heartbreaks, love interests, etc. I here wish to make statement which will, no doubt, NOT go down in History but is nevertheless true — Them wuz the Days!

Goodnight Mary. See you when the War's over. Luf, Joe T.

Young Cadet Joe at ease

Young Cadet Joe
striking a casual pose

Into the War

After the 27 October 1944 letter, Dad is not heard from until 21 January 1945, in a letter written from India. The letters are all addressed to Mom in Dayton, Ohio, which indicates she continued her work with the Veterans' Administration.

I know little about Mom's time in Dayton other than her rueful remark that she had imagined herself as a "career girl in New York City" but only made it as far as Dayton. While still in high school in Sullivan, Mom and her best friend Norma (*nee* Stinson, later Pirtle) had specific priorities for dates and boyfriends. Sons of farmers got extra points because farmers got a special allowance of otherwise rationed gasoline for their tractors, so their sons could drive on dates.

The Sullivan High School yearbook *Purple and Gold* for 1944 forecast this "prophecy" for the two seniors looking ahead to 1954: "Norma Stinson and Mary Jane Griffith are New York bachelor girls." Mom, of course, married Dad, and Norma married Johnny Pirtle, who appears in these letters as "that Pirtle guy." After Dad died and Mom moved to Prescott, Arizona, she was for several years a part-time neighbor of Johnny and Norma, who had a "snowbird" winter home just down the road from Mom's house. When Johnny died, Norma sold that house and stayed year-round in Ohio, close to her children.

Mom and Norma remained best friends for life. Even when they lived far apart and no longer traveled, they telephoned regularly. Mom passed away in Arizona in 2016; Norma died in 2018 in Ohio.

Dad told an interesting story about how he got to the war. This obviously took place not long after his 27 October letter. He was in C-47 training, somewhere near Memphis, Tennessee in his telling. I have not been able to identify what base that might have been. His SQR cites an "Aircraft Maintenance and Engineering School" in Memphis, but he was there in 1946, after his return from India. We have a photograph of Mom and Dad as newlyweds during an evening out, and they said it was taken in Memphis, so that tracks for 1946 but doesn't help for 1944. At least we know there was a USAAF facility in Memphis.

Anyway, wherever he was, one day there was a notice posted on the bulletin board. It said that anyone who wanted to go into the war right away should be on the tarmac the following morning at 0400. It was already October 1944, and Dad was afraid the war would end before he got there, so he and a friend showed up at the appointed time and place. There they found twenty-one new C-47s lined up, still factory-shiny silver. Dad became the pilot of one because he had six hours of C-47 time. His friend became his co-pilot because the friend had only two. Their navigator had never flown over water, but he inspired confidence, Dad said, because he was good at math. Off they went.

Their route, as best I can recall from Dad's story, was Memphis to Fort Benjamin Harrison (near Indianapolis); to upstate New York; to Bangor, Maine; to Goose Bay, Labrador; to Greenland; to Reykjavik, Iceland; to Great Britain. Of the twenty-one aircraft that began, Dad said seventeen arrived in Britain. Dad said he never learned what happened to the other four. The mission was considered a success. Remembering the stars over the Atlantic after they climbed above the clouds, Dad said, "I was 19 years old, in command of my own aircraft, off to the great war. It was the greatest moment of my life." He later modified that "greatest" remark to include his marriage and the births of his three children, but I think he told the truth the first time, and I don't begrudge him that.

Dad flew briefly in Britain and France (SQR says October, November and December), and then flew to Italy. He told me he passed through Cairo. From there he would have flown via the Middle East or the Horn of Africa. All I know for certain is that his next letter, dated 21 January 1945, is from India.

India
January—February 1945

21 January 1945

Hi Sweet and Lovely, Greetings from the wilds of Calcutta and vicinity. [This was Dum Dum, a British Royal Air Force base – RAF Dum Dum.] I'm ten thousand miles from Dayton and can't go much farther without getting closer. (I figured that one out scientifically – proud of me?) Confidentially I'd like to go a lot farther and get a little closer.

I haven't had any gossip from that part of Ohio in a helluva long time. In fact I haven't had a letter from anywhere since November. Sure is a rough war. I've decided I've been traveling around too fast for the mail to catch me. To date I've been on four continents, in fourteen countries, used 9 different kinds of money, listened to seven different languages (*!&xx&&**) (!) and before the summer is over I confidently expect 43 different kinds of malaria.

If you get down in this corner of the woods in the next year or so drop in for a spot of tea, old girl.

You'll find me living on the American side of a British airfield in one of the most modern tents equipped with hot and cold running mosquitoes, ants, rodents, reptiles, etc.

If I had this place back in the states I'd call it a zoo and charge admission.

The Japs are only a couple hundred miles away and the smell is terrific when the wind is in the East.

I trust your heart hasn't been stolen away by any of those handsome Air Corp wolv - er - men at Wright Field.

Since I haven't heard from you in so long I'll send this letter to Sullivan because I'm not sure you're still in Dayton. Anyway, however and wherever and whenever you get this letter, remember that on the night of January 21, 1945 at 9:03 P.M. there was a guy somewhere thinking of you and saying Goodnite, Love, Joe T.

Dad's next letter is dated 13 February 1945 at 6:35 PM and is from Calcutta. In it he recounts some of his itinerary in detail.

13 February 1945

Hi little ol' gal – Evening in India – blue evening. Have a table out in front of my tent. It's very still and cool and down the line I can hear someone's radio playing "Will I ever find ___" [I believe this was "My Ideal" by Frank Sinatra.] I'm going to have to hurry because it's getting dark and I'm going flying later tonight. Your letter is about thirty minutes old and after going through it three times I detected for the first time a serious note. You must be growing up little girl! I guess we all are. It scares me sometimes.

This is an awfully nice snapshot you sent. You look very much like a lady and not like a girl anymore. It's dark and I'm going to have to move inside.

2 minutes later – Hello again you. You asked where I've been so I'll tell you as much as I can. I left a field in Maine one sunny day in October and went to Labrador. From Labrador to Iceland to Scotland to England.

Here he omits Greenland, between Labrador and Iceland, which I have a clear recollection of him naming as a stop on that route. Whether the omission in his letter was for security reasons or was just forgetfulness I don't know.

Stayed in England about a month during which I made trips to North Ireland and numerous other trips to France. My Squadron moved to France and we lived in Paris for about 6 weeks flying out of Le Bourget Airdrome.

Dad's address book includes the entry: Miss Eity Proszoroski, 9 Rue Reaumer, Paris (3d) France. Another Paris connection was that Edith Piaf remained a favorite singer of Dad's to the end of his life.

I was ordered back to England and we all thought then we were going back to the States for special training so there was much rejoicing till the Captain came down on the afternoon of Christmas Eve and told us we were going to China-Burma-India. If you've ever been just a little homesick you'll know that was quite a blow so we had a party at the club that nite and when I left a few days later they still hadn't repaired all the damage to the building. [With Germany nearly finished, air transport priority shifted to increasing the tonnage delivered to China from India.]

I spent 4 nites in London and then went on to Naples. Saw the Isle of Capri and visited the ruins of Pompeii. Spent a night on Malta and on to Cairo. Saw the Great Pyramids and the Sphinx there. The Sphinx was much smaller than I had imagined and the surprising thing about the pyramids was the fact that we were allowed to go inside and see the tomb of some ancient Egyptian king whose name I didn't trouble to find out. That's about it. Now I'm a few miles from Calcutta. Saw the Taj Mahal at Agra from the air on the way across India. Took a little hop down toward where the flying fishes play one day.

Dad provided no details on his flight path from Cairo to Calcutta, possibly because that information was classified. Air way stations on the route to India included: Asmara Airport in Eritrea, RAF Khormaksar in Aden; RAF Lydda in Palestine; RAF Habbaniya in Iraq; RAF Muharraq in Bahrain; and RAF Sharjah in Oman. The likely entry point in India was Karachi, now in Pakistan. I do not know specifically to or through which way station(s) he would have flown.

I'm flying a transport, naturally. Life isn't too dull. There's a swimming club with a nice indoor and outdoor pool and a bar so I don't spend much time in the water, naturally. Best of all it has a high stone fence around it that shuts out the dirty city of Calcutta. You can ride anywhere in town in a rickshaw for a rupee (about 30 cents) and you see snake charmers sitting on the sidewalk with baskets of cobras which they'll make dance for a rupee, little native boys selling everything from parrots to monkeys to hot tips on the horse races of which there is one every day. 90% of the female population are of the dark skinned variety and the other 10% are British officers' wives so don't you worry, I'm being a good little boy. Gotta go now. Be back in the morning – I hope.

This letter resumes the next day.

Feb. 14, 4:30 P.M.

Hello you gorgeous Creature – Just got out of bed a short time ago, went to the mess hall to correct my famished condition and here I is back again. Looking back I see that I was busily telling you of the highlights of Calcutta called. Anywho –

Mary Jane Griffith 1945

The natives never bury their dead. The bodies are all cremated at an open public oven in a section of Calcutta. If the family is too poor to buy the expensive sandalwood used the bodies are thrown into the river.

There is one very nice Cabaret Restaurant in town where one may go for a cool drink and to dance if one has someone to dance with. In most cases one hasn't, unfortunately.

I'm sending, under separate cover, a picture we had taken in Egypt. The big guy on the little donkey was our guide and all the rest of those guys' names are Joe. That pyramid, incidentally is the largest of nine.

You'll notice that the Sphinx is reinforced with sandbags (under the chin) to protect it against a possible air raid. That pile of rocks on the left is somebody's tomb. I couldn't spell his name anyway.

I was very much interested and surprised to learn that I'm getting married. This must be some propaganda put out by some so-called friends of mine. It burns me up, particularly when I'm 10,000 miles from home and have no serious intentions of marriage for numerous years. Betty Jo is a swell girl and I'll tell it to the world but then marriage is another thing. There are no strings tied to me yet so please go on eating yore poor lil' ol' heart out for me, 'cause I love it.

This is clearly related to some social intrigue about which I have no further information. From a note on the back of a photograph, I believe the reference was to Betty Jo Knotts, who seems to have been a girlfriend of Dad's. She did not attend Sullivan High School, so she is a candidate to have been the "Illinois woman" who may have expected or hoped to marry Dad.

> I know now I want to be home before the Spring of '46. Damit. And you'd better be waitin' and not with a handshake!
>
> Gertrude is probably thinking all sorts of foul things about me because I haven't written but gee I'd love to be back at S.H.S. jus' whistlin'. And throwing starch at angry little girls on the bus. And running a Pony Express. And High School Proms. And basketball games. And bridges over the water. And a coke at Frew's. And bus drivers named Herman. An' – But that's enough for now. Before I wind this up I'd like to say that this here sho' is a swell snapshot honeychile and – How about some more? Be good and do right and write. Love, **Joe T.**

Next is a V-Mail postmarked 27 February 1945. According to the National WWII Museum, V-mail, short for "Victory mail," was a postal system put into place during the war to drastically reduce the space needed to transport mail thus freeing up room for other valuable supplies. "A special form is used which permits the letter to be photographed in microfilm. The small film is transported and then reproduced and delivered. Use of V-mail is urged because it greatly furthers the war effort by saving shipping and airplane space."

> **27 February 1945**
>
> Dear Mary, Just between you and I and the guys who photograph this letter you owe me a letter but since I was thinking of you anyway I may as well write and tell you what guys ten thousand miles from home think of in their spare time – Pretty girls like you, naturally. With inspiration like that we can't lose. Just give me a little time to get started and I'll have this war finished up in no time and be back home throwing starch at you again.
>
> Oh, so you take dictation from a chair, alone, eh? How would you like to be my secretary? "Oh, please take a letter Miss Griffith – Just address it F/O J.T. Pound\F.B.I. (Forgotten B– of India). Luf, **Joe**

Mom's job at the VA in Dayton was as a secretary.

WHEATLEY (WHEATLEY) PUMPER

"Acme Business College? Remember Miss Stanton, the student you selected as most likely to succeed? Well she has !"

Wheatley Pumper cartoon - see Dad's handwritten note

India

March 1945

Dad's next letter is dated March 8, 1945, from "India - Damit!" He opens ebulliently:

March 8, 1945

Hello you great-big-gorgeous-armful-of-woman-you! All I can say is it's about time I had a letter from you! Every day I dash madly up to the post office – but no letter from Mary – until today! I've been in bed all afternoon – recuperating. Gee, I wish I could spend more afternoons in bed!

Say that new hat of yours must be a honey! But then on you even a dish of potato salad would look good. (This is going to cost you money, woman.) [This last statement must relate to an unknown exchange.]

And by the way, I distinctly remember writing you at least two letters when I was in France. Dincha get any of them? [I do not have those letters, so Mom may have never received them.]

Life grows duller and duller. No one to fight with but a couple million mosquitoes. I give up – I'm outnumbered. They aren't interesting to fight with anyway. They can't even throw starch or pull hair.

By the way – you didn't get another one of those haircuts recently did you? I remember you used to get a wild idea about once a year and come to school looking like a fugitive from a broom factory. Omigosh it was horrible! I don't see how I ever carried on under the strain!

Incidentally I wrote you a letter and told you all about my journey across the world and sent a picture. Didya get 'em? By the way – Columbus was right – it's round.

Hey, you hadn't better be a peroxide blonde! I distrust blondes and redheads – (and most brunettes).

I have 5 gray hairs already worrying whether you're going to break down and write to me so don't get excited about those 2 you have.

I make a journey into darkest Burma now and then. The women up there are all brunettes – face and all. This Calcutta place is a helluva town. The liquor is terrible.

My honorable tent mate is snoring something terrible. Dear me I don't know what I'm going to do with that character! Every night I have to wake him up and then try and beat him to sleep.

Have you heard any more good dirt about me? I'm always interested in what my friends are going to do with my life.

[The text on the last page is written in two connected spirals.]
If this letter is below my usual standards it's all your fault. You made me dizzy and if you'll please excuse my abruptness I might say in more ways than one you'll have to figure that one out for yourself, little girl. This is a very crazy way to write a letter don't you think? But then it would be a very dull world if everyone were sane wouldn't it? I think people have more fun than anybody. [?] I'll wind this up for now so good night. Luf, **Joe T.**

March 10, 1945

Hello Sweet Thing. When I received one letter from you I considered myself a very lucky little boy indeed. But when I get three in a row on three consecutive days – well, I'm speechless Miss Griffith!

Anyway it's a very pleasant feeling and I don't mind at all. For your information the latest was postmarked February 28. That, little girl, is very good service for India. Ten days!

I'm going to be interrupted right in the middle of things I'm afraid for two friends are surreptitiously conspiring to drag me off to the movies. We have one of those open-air affairs but sometimes some fairly good pictures. Tonite it's "Meet Me in St. Louis." I can think of better places to meet Judy Garland but that's neither here nor there.

Personally I think if you want to join the Catholic Church it's nobody's business but your own. You're intelligent enough to know what you want to do so you just go ahead and do what you like and don't let any old hens talk you out of it. People who meddle in other people's affairs make me mad.

This paragraph was a surprise. Mom seems to have raised in an earlier letter the prospect of converting to Catholicism. Apparently, the attraction passed; I don't recall her ever mentioning this to me. Mom's religious views could be characterized as agnostic in an open-minded way.

Don't send any more letters to this address. You'll be hearing from me in a few days and I'll give you a new bunch of figures to send your letters to. I'm going up country to fly the Hump. Just when I had the Calcutta swimming club half paid for, too.

Apparently, Dad went to the movie at this point, and then resumed.

Hi you – "You will be my tootsie wootzie (or something)" [illustrated by musical notes]. How about meeting me in St. Louis? Say about – umm – May 1946?

Confidentially I've seen better movies – and my only memory of St. Louis was a smoky railroad station and three minutes between trains. But it's U.S.A. so I'll take it.

This will be all for tonite. I'll write again soon. Of course where I'm going the mail may have to come out by pack mule. Goodnight – Love, Joe T.

March 11, 1:00 A.M., P.S. I was just remembering a dance we went to once. Remember? it took me three days to get nerve enough to ask you for I didn't think you would go with me – I being a moron. Didn't we have fun though? And I fell over you, me and my feet.

Do you remember when my dog ran away and went to your house and I had to go get him? I never could get that darn dog to run away and go over there again!

If you're wondering what I'm doing up so late – I never do go to bed until about this time. We've been having quite a bull session but everyone else has finally gone to bed. They would no doubt appreciate it if I'd get this light out.

I liked that part of your letter where you said since I wasn't there you blamed that starch business on me naturally. Woman, I'll fix you when I get home.

Incidentally, I'll have a 30 day rest leave when I get back to the States. No less than 3 different people have all decided already just what I'm going to do with that 30 days. I'm afraid I'm going to make somebody mad.

Yep, it's been nearly two years. I wonder if you can still lift that right eyebrow and look oh so very cynical. (I just had to wake Gardner and ask how to spell that word.) [arrow to "cynical"] Guess I'm just a low brow. Gardner says, quote, "27!?xx^&! Whyinhell don'tcha buy yourself a @xx-!& dictionary &@!&!" unquote, paragraph.

Doug hasn't written me in nearly a year now so you're lucky to get a one page letter. He didn't answer my last three letters so I guess I must have done something wrong. My folks are living in Terre Haute now and he was down there one weekend around the first of the year and stayed all night.

This P.S. is longer than the letter now so I'd better douse the glim and get in the sack for some shuteye.

March 13, 1945, 1332nd AAF Base Unit

Dear Mary — Jus' a note to say this India place shouldn't happen to a dog. Then too, I wanted to give you this new address so maybe you could be persuaded to write me a letter now and then.

So you've taken to drink? (!) Tsk, tsk! Mary, you're the last girl I would have suspected would be hoarding alcohol in her room! But you're more or less unpredictable, being a woman and all. The best way to try and understand women is not to.

This little boy has had a busy day so I'll be saying 'Goodnight' for now. Yours always, Joe T. [I conclude from this letter that Dad had indeed moved up country.]

March 18, 1945, Sunday afternoon

Hello Mary. I've been pacing up and down and listening to the drip, drip, drip of the rain and damning such weather and so decided I'd better settle down and do something sensible and interesting such as writing you before I blow my top.

Of course my doing something sensible is probably beyond anything comprehensible to you but we aren't here to discuss the sanity, or lack of it, on my part. Being a woman you would probably say that I, being of the male of the species, wasn't supposed to have any brains anyway. For the time being we'll accept your viewpoint until I have time to convince you otherwise.

So now let's talk about the weather. No! No! Not that! Incidentally it's stopped raining but just you wait a couple minutes. All the mosquitoes around here are installing water wings.

A strange thing happened last night. Yep. I saw a white woman! Some nurses came down so we had a dance at the club. Of course they were outnumbered about eight to one so they fought a losing battle. The liquor supply was ample so it got pretty drunk out. My pal, Steve, was about to stall out around eleven o'clock so I had to convoy him home before he spun in. [This may have been Steve Pound. He is discussed in detail later.]

Outside of that there hasn't been anything interesting happening around here. Much. . .

It's started to rain again. . . .

I just received a clipping from my mother that Bertha Able's brother was killed. The letter was mailed October 16. What ever happened to Bertha?

There are a couple of characters sitting here discussing – of all things – duck hunting. I give up. How can I think? I can't win. So goodbye for now. – Write. Love, Joe T.

Dad's Base—Mohanbari, India

April 1945

The letters from April include a reference to what seems to have been an in-theater Rest and Recuperation break. The first letter certainly written in April follows.

April 12, 1945—1332 Base Unit

Hello Mary. Maybe it's a bit late to be writing a letter but then I always write my best letters around midnite.

I've just come back from a four day rest at the jungle camp. We paddled up and down the Ganges in an Indian dugout they had out there – went swimming and fishing and hunting and lay in the sun. Had a pretty good time. Last nite one of the boys came in with 3 wild chickens so the cook fried 'em and made coffee and we sat in the mess tent and listened to the rain.

Yeah, I know, I don't like this paper too. But it's all I have so you'll have to suffer along with me. I only go to the PX when I run out of cigarettes and beer. All I need now is some passionate poiple (sic) ink to go with this blue.

Where's your brother now? Is he still in France? [Mom's older brother Wayne was in the army in Europe.]

The liquor situation here is in general pretty good. We won't talk about the food other than to say it runs mostly to dehydrated potatoes. I don't want to unload all my troubles on you – just most of them.

It's midnite here so it's probably about 7:30 in Dayton. You're probably just waking up and looking horrible, naturally.

I had a nice long letter from Gertrude so maybe she isn't too mad at me. Next time you go home will you please check and see if that bridge is still over the water?

I hope by now you have a roof over your pretty little head. I didn't like the idea of you having to pitch a tent in the middle of Main St.

I'll send you some pictures of my happy home as soon as I can borrow a camera to fit my roll of film.

Bill is trying to sleep and I'm not helping much with this light on. He started over the Hump tonite but came back in a hurry. These airplanes don't fly so good on one engine, particularly with a load on. Goodnight now, Yours always,
Joe T.

The following letter is dated the next day.

April 13

Dear Mary. I get these letter writing urges in bunches. Probably before long I'll get over it and won't write you for three or four whole days at a time. It's all your fault – you won't write to me unless I make with the correspondence like mad and keep reminding you.

The writing will be unusually unreadable tonight. Bob is using the table so I'm making like my knee is a desk. You'd be surprised what one can do without and pretend he has. It makes for a happy existence.

I wanted to remind you to see a picture called "Destination – Burma" with Errol Flynn. At the beginning of the picture make a mental note of the name of the little town marked on the map up in the corner of India where the P-38 lands and some day I'll tell you something about it that might interest you. And, too, you'll get an idea of the kind of country I have to walk back thru if I ever have to walk back.

The Japanese aren't much of a problem anymore. They've had their day.

I slept all day (oh luxury!) so I'm being mean and keeping every one else awake with the light on. They don't complain – much. We have a very unusual bunch of names here – Bob, Bill, John, and Joe. My name is Joe. Hi Mary. I think I'll go over and see if I can talk the cook out of something to eat. So goodnight now.

The next letter begins "Somewhere in India, Sometime in April (I think it's about the 15th or 16th)."

April 15-16

Hi Sweet Thing. A mere man has two strikes on him from the beginning.

The minute I turn my back all my oldest and truest girlfriends go galavantin' around all over the state of Ohio missing buses and seeing double features (it sez here) until the wee hours of the morning! Tsk Tsk! Excuse me for a moment while I go over in the corner and shed a tear for a lost cause.

The character next door has been playing his blasted portable phonograph again. For the evening's hilarious entertainment he has picked such joyous selections as "I'll Walk Alone" and "When You're a Long Long Way From Home" and stuff like that. He has, indeed, helped the morale of all troops in the neighborhood to reach a new high in lows.

Perhaps if you were here you might even prevail upon me to kiss you tonite – if you twisted my arm hard enough. I remember you kissed me once. You've probably forgotten but I was just getting ready to tie on a big drunk and you were trying to persuade me not to. Confidentially, I've always suspected you did it to make Doug envious but don't tell me the truth, please, – jus' let me go on dreamin'. [This is almost certainly a reference to Dad's friend, Doug Pace.]

It's just after dark and all the mosquitoes are busy going around giving malaria to the bad little boys who don't take their atabrine every day. They used to send the boys back to the States when they had malaria but they don't do that anymore so there's really no point in catching it.

I know a fellow who has been over here seventeen months. He walks in his sleep and raves at the moon. And goes around all day mumbling something about taking out citizenship papers here when the war is over and stealing away some savage maiden from a native chieftain he dreamed up between rains.

By the way, young lady! Don't you know Esquire is a man's magazine? How do you expect me to make beautiful music if you go around stealing my trade secrets?

You're a very rare specimen. You're one of the few females I ever knew who showed any indication of having any intelligence. Now don't go and get stuck up – notice I merely said "an indication." I still think it's impossible.

I have a very strong feeling that that little man is going to be around shining that light in my eyes before long so I'd better say Goodnight for now. Keep smiling, dear Mary. Love and Stuff, Joe T

April 23, 1945

Hi Little Girl. In view of the fact that I'm going to be kicked out in the cold by one of my very best friends unless I dispatch a letter P.D.Q. (it sez here) I hurriedly brushed the dust off the trusty old Parker and took same in hand to – but this already sounds too stiff and formal to be a letter of mine – I'm just a lowbrow – so let's forget the corollaries and adverbs and just say "Oboy! Wuz I happy to get two (2) letters from you today!"

So I beg of you to please not set the expiration date on that lovin mentioned just yet. Set it up to sometime in 1946 and by that time I'll be home to protect my interests.

I noticed that part of one letter was written April 4, part April 9, and the whole business postmarked the 12th. How come? Please accept these stamps I'm enclosing under conditions that you mail your letters the same day you write them and thus save much wear and tear on my fingernails, nervous system etc.

On account of the &!^szc?! climate around here those stamps were all stuck to the oil paper on the back but I thoughtfully pulled 'em all loose for you – all but one, just so you can see what I suffer. The damp climate is one of the pet gripes around here. It's almost impossible to keep envelopes – they all stick together and make a helluva mess. Clothes mildew, shoes fall apart etc. There are probably lots of nice things one could say about India but I can't seem to think of any just now.

When I went to the jungle camp I did see hundreds of beautiful wild orchids. The only comment I have to make concerning orchids is that they would probably be at their best with you and a black velvet dress. – Maybe we'll see.

Mary Reed finally broke down and wrote me a very nice letter which same I received about two weeks ago. [Mary Reed later married Bill Able. Both were good friends of my parents whom I knew as boy.] I was very much puzzled when down at the bottom of the letter someone added a couple remarks about me not answering her last letter and all that and signed it "Norma."

I couldn't figure out who I knew named Norma who ever wrote me a letter. So finally I decided Mary's younger sister's name must be Norma (by the way what is it?). I hazily remembered that she sent me a card some 2 years ago. So with the proper big brother spirit I sit down and write a nice little letter and address it to Norma Reed. Stamp and all.

Then all kinds of doubts assail me so I tear up the letter and, to be on the safe side, tack a note on Mary's letter and tell Norma if she'll write me another letter I'll promise to answer it.

Damit I should have thought of Norma Stinson but I supposed she had forgotten me long ago. Anyway I received a nice letter from her today.

To say the least I felt very silly. Yeah, I know, I'm a bird brain. But it must have been nearly a year since I didn't answer Norma's last letter. So how was I to know? Anyway don't say anything about this and I'll write her tonite and see if I can get out of the doghouse.

By the way – I wrote Gertrude and told her I was still dreaming of her Betty Grable legs. I must have been a troublesome kid in those days.

You know my album has only one snapshot of you. You must have had more than that taken in the past two years. Why don't you break down and send a couple to help the Hump jockeys thru the monsoon season?

It hasn't rained for two days now. That's something worthy of a place in my diary.

You must have had at least a couple letters from me by now. Anyway don't you worry your pretty little head about me. I don't want you to get wrinkled and gray you know.

My roommates have already decided how they're going to divide up my shirts.

But if all goes well I might make it home by Christmas. By the way – do you know where you can lay hands on about 2 truckloads of mistletoe?

I'd better wind this up for now – so be sweet and keep smilin'. Yours always, Joe T

April 25

Hi Mary. This is a particularly dreary corner of the world so this may not be a very cheery note – not necessarily because I'm blue but just because of circumstances – there jus' ain't much to be happy about.

I don't know how I happen to be sending you two letters in three days. Maybe I'm just naturally susceptible to pretty little girls with brown eyes who have a bad habit of tossing good liquor over the back fence.

I went aviatin' to China yesterday. Got back a bit late so slept most of the day. I can think of much more enjoyable ways to waste my young manhood. Oh well, I probably wouldn't have done anything worthwhile with it anyway –

I was in a big rush to get outside the continental limits of the United States for there was an Illinois woman hot on my trail. Now I'm afraid to come home. Just an exile – By the way – any reminiscing I do with you concerning my past love life is of a personal nature. I'm probably making a mistake confiding in a woman. But then you've been one of my most trusted confederates as far back as I can remember. I can even remember when you were chief messenger of Pony Express Ltd.

Remember that hectic two week period when I was madly in love with Norma Keller? Stinson tells me that old female is still around Sullivan. I supposed she had been run out of town by now.

I didn't have much faith that this was going to be a very intelligent letter when I started but it's the only way I can get you to write me so I'll send it and let you suffer. Besides it's chow time. So G'bye now. Write. Love, Joe T

May 1945

Included among the May letters is one that belongs in that month by virtue of the context, and dates and events mentioned, but is dated 12 April, clearly in error.

May 5 Same damn Place

Dear Mary. This is a particularly quiet and peaceful afternoon. El Kapitan Gillette is solemnly playing solitaire and Leftenant Villarosa (Bill to you) finally decided to go shave – after all these weeks. I'm just being lazy and loving it.

Then, too, the sky has a pretty, rose colored glow today because the mail plane brought me a letter from a nice young lady in that there furrin country some folks calls the United States. In case you haven't figured that one out I was referring to one Mademoiselle M. Griffith, who specializes in bright, cheery letters extraordinaire (or sumpin').

With no work to worry about, the rain stopped, and my shoes off I can just relax and wriggle my toes and consider all the reasons why I shouldn't be pleased with Life and stuff. I journeyed a muddy several miles into the village this morning for a new pair of boots. Everyone here wears Natal boots. They are made of soft leather and come halfway to the knee and protect the ankle from malaria packin' mosquitoes between sundown and dawn.

Same day, 9:00 P.M

Hi again – First it was chow time and then Bill dragged me off to the movies so let's take it from there.

We can always gripe about the food. It isn't too bad, really, considering the supply problem. Everything we get comes from the States except a few items such as fresh tomatoes, onions, cucumbers, etc. which the mess officer goes out and barters for. How did we get talking about food?

Why don't you get this Garb-ish [an apparent reference to clothing] mood oftener? It rated me six pages this time.

By the way – if sometime you don't get a letter for a few weeks or the War Dept. notifies my mama I didn't get back for the evening meal don't rush out and buy a black dress right away.

I know one chap who came walking down out of the hills two months after everyone had decided he had gone to join his honorable ancestors. This little clipping enclosed relates a few experiences of some fellows I know. I fly over this country every trip. [The clipping, cut from what looks to have been the unit's newsletter, is headlined:

NAGA HEADHUNTERS WELCOME FLYERS

WEIRD ADVENTURES FOR DELONGA CREW]

And, too, I'm sending a few lousy snapshots I picked up here and there. They're not very good snaps and the negatives were printed here in India so that partially explains it. Anyway, you might be interested in the blurred scenes of some of my misadventures, etc.

Next morning. I wandered off last nite but I'm back now so we'll continue with the next installment.

You know if you keep working hard I believe you'll be running the Veterans' Administration in a couple more years. Maybe when the war is over I'll be around for a job washing windows or something.

You're probably the first pretty girl with brains they've seen in a long time. I think you're doing pretty swell but I hope you don't turn into one of them there career girls. Somewhere a long time ago I read that a career is O.K. – but you can't run your fingers thru its hair.

By the way, anytime you have to let off steam I'll be the willing recipient. I rather enjoy it. Ho hum. That poor Pirtle guy! Tsk, tsk.

You're a mean wicked little girl, Mary, but never mind – you just stay that way –

Oh! so I'm a brute now? I'll fix you when I get home.

Since you have a couple of new dresses I've decided I may as well have a new picture of you. This other one is getting a bit travel weary after some 30,000 miles of snow and rain and muddy airfields. Of course it's merely the dress I'm interested in. I didn't realize it before but I've just decided that I'm deeply interested in -er- dress design. Ahem. It might be a good idea if you were in the dress, of course. . .

Anyway, that's what the check is for. I can't find any place to spend my money and I don't like the miserish feeling I get when I save all of it.

I want a Sepia finish and a leather folder if you can find one. Anyway something durable. And send it first class so I'll get it before next Easter.

I just remembered that today is Sunday. Yeah, I know, the Devil has his hand on my shoulder.

Since this is the last sheet of paper between here and the PX I'd better be thinking of a suitable way to wind this up. Lemme see. My mind is blank so I'll just say g'bye for now. Love etc., Joe T

May 9

Dear Mary. If it wasn't for the fact that I'm going to have to fly a little later I'd go over to the club and get stinkin' drunk and forget all – but since these unfortunate conditions prevail I'll just unload my blues and stuff on you – again –.

Gee, I'm glad you have nice soft shoulders. Do you mind if I use one for a while? I'm disgusted with things in general but since I'm in the Army and a lowly F/O (that's two ranks below a garden worm). I just grit these three teeth I have left and maintain a burning silence.

An F/O was a Flight Officer. Flight Officer was a United States Army Air Forces rank used during World War II, from 1942 to 1945; the rank being created on 10 September 1942. Enlisted and aviation cadet trainees who successfully passed air qualification training were appointed as Flight Officers and served as rated pilots, navigators, flight engineers, bombardiers and glider pilots.

At the end of World War II, the Army Air Forces discontinued using the rank of Flight Officer. By then, all the service's flight officers had either been promoted to commissioned officer ranks during the course of the war or discharged.

May 9 (continued)

Then, too, this character next door insists on playing such things as "Moonlight Mood" etc. on his blasted phonograph. It's times like these that try men's souls.

I was just thinking about such things as the old bowling alley out west of town, and the old gang, and the fun we had, and high school Proms, and Wentz's after a basketball game, and girls named Mary –

Hi You – next afternoon. I was rudely interrupted midst all those reminiscent dreams so had to lay aside my opium pipe and go to work.

Anyway, I've been to China and back and the fact that I haven't slept for two days might have something to do with the dullness of this letter. I wanted to finish it before I go to sleep for I probably won't come out from under the red haze before tomorrow afternoon. And just to show you how appreciative I am (no cracks!) I wanted to answer this letter of April 30th I had here when I got back.

You can put down that hat pin for I'm over that dark brown mood now. I'm too tired to fight. Tsk, tsk! The names I do get called! Last letter I was a brute – now I'm a heel. tsk, tsk! You'll have me believing this stuff myself!

The liquor situation is pretty good. We get a quart of State side whiskey every month and a case of beer. Then we get 2 ozs. for every trip. I let mine accumulate for 16 trips and that makes another quart about every month.

And, too, we have a bar in the club that never runs dry. Of course I always give away my liquor ration to my friends as soon as I get it. (My best friend is me.)

I'd better get this letter in the mail before I drop off to sleep. G'bye now. Love always, Joe T.

Despite its April date, this next letter was certainly written in May. From its context my guess is 12 May, but that's only a guess.

April 12, Afternoon

Dear Mary, Since I'm scheduled to go out with the next

airplane that comes in I've decided to answer your half-hour-old letter now because I'll be awfully tired and sleepy when I get back tomorrow.

This new Stateside paper came in a package from my Mama this morning. Apparently it is a gentle hint that my semi-monthly letters aren't enough. I think she's finally convinced herself that her baby boy is growing up. Anyway she included a tube of shaving cream. If you don't think I'm the sweetest boy in the world just ask my mother – she'll tell you. Ho hum. I'm a spoiled brat.

The biggest news today is that it isn't raining. The end of the war in Europe is getting only secondary consideration. Special Services has been saturating us with War Dept. propaganda films since V-E Day. The moral to most of the pictures is: "Just because the war is half over don't get careless, bub." It's getting a bit old and trying on the nerves by now.

The letter is dated 12 April but V-E Day was not until 8 May.

Since Rangoon has been taken the old Burma road will probably be in operation again in a short time. And with the Ledo Road already open there may be a few Hump jockeys out of a job one day.

The British and Indians took Rangoon from the Japanese in early May, another indication that this letter is misdated.

El Capitan Gillette is peacefully sleeping. He's finished his tour and is waiting orders to go home. Back to that furrin country – the U.S.A.

By the way – don't be surprized if you find a package in your mailbox about the 27th of next October. (I just mailed it this morning.) It's something I picked up in my travels that probably won't go so good with brown eyes but I'm not very experienced along that line. Anyway I liked the color.

And incidentally – you don't need blond hair and big saucer blue eyes to look innocent – just you hang on to that flexible right eyebrow –. Besides I detest blondes. That is – some blondes –.

I'm sending an "Esquire-ish" clipping to end all clippings. Personally I think I shouldn't send it but this is the best I've seen and I've seen some honeys. Apparently this method of advertising sells numerous quantities of pumps and valves.

I gotta go now. Damit. Be good and write. Love etc., Joe T.

The foregoing description may relate to an undated cartoon from the *Wheatley Pumper* that was among the collection of letters. It shows an attractive young woman sitting on an older and well-dressed (and bald) man's lap while talking on the telephone (see page 28). His face has lipstick smears on it. The caption is: "Acme Business College? Remember Miss Stanton, the student you selected as most likely to succeed? Well she has!" And a handwritten comment — in Dad's handwriting, I think — reads: "I very diplomatically refrain from comment."

May 18

Dear Woman. My mama sent me this paper together with instructions that she was to get it all back – one piece at a time with writing on it. I'm cheating on her just a little but since she doesn't know about it I'm sure she won't mind.

I've just been browzing thru some old letters of yours of which I have thirteen. I wish you would hurry up and send me another one so I can get out of this thirteen rut. Every once in a while I get a queer and unusual feeling of sanity is I just dig out a couple of your old letters and it generally puts me back into my usual unbalanced mental state. I'm going to be awfully uneasy until I get No. 14 tho.

I spent three days in China waiting to have a busted part on my airplane fixed. India is hardly the garden spot of the world but it sure seems heavenly after China.

We got Gillette off for the U.S.A. this afternoon.

That would presumably refer to Captain Gillette, the El Kapitan Gillette mentioned in earlier letters.

Bob is trying to entice me over to the club with reminiscences (or something) of a cold beer. But I'm staying to finish my letter like a good little boy. Hey! What did I say? Cold beer? I gotta rush off now honey chile. Be good and write. Love etc. Joe T.

May 20

Dear Woman. I should be getting some sleep but I can't sleep very well until I answer nice letters like that one today. Incidentally it's about time you wrote!

It's raining again and believe it or not I'm glad – it's been so damn hot around here for a couple of days. t's nice and cool now and I have my shirt off and haven't seen a mosquito tonite. That is unusual. I've been donating plasma to half the mosquitoes in Assam.

It would be a very peaceful evening if I didn't have to fly – and if I weren't a little lonely. – Bill's pin-up pictures are awfully poor company. Why don't you come on over?

Holy Cow! Now the truth comes out! You were madly in love with me in the sixth grade! Women are beyond me. In them days I never suspected that as far as you were concerned I was anything more than dirt underfoot. And when you turned that icy look on me I always had to fight down an almost irresistible desire to hide under the seat! And you used to be always throwing paper and pestering me all the time! Oh me. I'll never be able to figure women out.

By the way, my mama sent me a clipping that Bill Faught is coming home. However I don't think it necessary that you go home to see him. Not that it's any of my business but I'm just kind of interested you might say – .

Bill just came in from a trip and he doesn't promise me anything good in the way of weather. I've got 21 trips over the Hump now and I've seen 21 different weather conditions – none of them too good.

I've got Bill settled down with a newspaper now so he'll shut up and let me write my letter. Have more trouble with that guy.

Lemme see. We were down to paragraph three in your letter.

The only way I can figure to get out of my uncomfortable situation without getting my neck broken is to take out citizenship papers and stay here when the war is over. But when I consider my great love for Coca Colas, strawberry milk shakes and such stuff as that one doesn't find in India I think it would be wiser for me to return to the Fatherland and hide in the hills and let you perteck me from this wily female [unknown].

I think I'll write to Herman and Doris tomorrow nite and set them straight on that starch deal. I'm finally going to tell the truth and confess that it was all your fault.

Who was the poor guy Bernadine caught with her little lasso? Some poor returning veteran who has been blinded by shell fire no doubt. Must have been shell shocked too to marry that woman.

Honestly, Mary, I believe I'm getting catty. I never did have any particular affection for that Pinkston woman though.

It was a reciprocal trade agreement.

There was a wench named Marjorie Brown who probably wouldn't have wept at my funeral either.

Bernadine Pinkston and Marjory Brown are in the Sullivan High School *Purple and Gold* yearbook as members of the Class of 1944, the same as Mom.

Gee, Mary, this is more gossiping than I've done in years and years. Sort of fun isn't it? Remembering old not-so-friendly-ships.

I suppose you know that the Army is going to discharge some men now that Germany is out of the war. I had to go down and fill out a form the other day. I told them I wanted to stay in till the war was over. It didn't make any difference whether I said yes or no for I'm going to be one of the last out since I drive a glorified furniture van but it was damn nice of them to ask me.

When I went to the Army I had it all figured out – I was going to grab the first available P-38 and destroy the honorable Japanese air force single handed. But there seem to be a bunch of guys who are doing a pretty good job of it without my help.

If I write much more my limited mental facilities won't recover sufficiently to turn out another letter for weeks and weeks so I'll say goodnight for now and run off to bed. Remember to write and I'll B.C.N.U. Love Always, Joe T

May 22

Dear Woman. I tried to type a letter on that old beat up piece of junk down at the briefing office tonite but when I got it home I couldn't even decipher it myself. – It was even worse than my handwriting (shuddup!!!) so I'm starting all over again.

Bill is in China and Bob is on temporary duty at another field for ten days and I can't even find anyone to swap lies with me. It's a particularly dismal evening (raining, damit!) and I'm in one of those dark blue moods.

I had a perfectly wonderful time today doing what I like to do best – nothing. I'm not really lazy, Mary, just tired you know. . . Hmmm . . . However this inactivity is getting on my nerves.

> The weather over the Hump is getting particularly obnoxious. On my last trip I was on instruments for seven straight hours and when I finally got on the ground in China I had enough gasoline in the tanks for about five more minutes. That was a most unenjoyable ride.

Mid-May was the usual start of the monsoon season, running to October. The monsoon brought turbulence, extraordinary rain, thunderstorms, hail, zero visibility and icing.

> You can keep right on writing me and saying what you want to. I rather enjoy it. And I won't think you're eccentric. At least no more eccentric than most women. Of course I'm prejudiced – being a mere male.
>
> Since we mentioned women it reminds that I'd better write to that Illinois woman some time before the end of the month. She gets a bit burned up if I don't write at least once a month. She threatens broken bones and fractured sculls (sic) and all sorts of fantastic impediments to my health but I just smile smug and safe for I'm a bit out of range, even for her, here in India.
>
> I finally borrowed a camera and took some pictures over the Hump the other day. If they are any good, which I doubt, I'll send you some.
>
> How often do you go home? Of course it's none of my business but I don't want you exposed to that Pirtle guy too much – you might change your mind about him. *You're* a bit out of *my* range, too, over there in Dayton, you know.
>
> I'm getting a little tired of this place. This is about as long as I've stayed anywhere for the past two years. I really should be happy and let good enough alone. The guys in China *really* have it rough – *they* don't get any liquor.
>
> And, too, it's just about exactly two years since I've seen you. I'm getting a little tired of that too.
>
> Yeah, I know, it's all my fault but I wish to state here and now that I was a victim of circumstances. Damn, I haven't had a good fight in two years.
>
> Do you have to take your vacation a certain time every year or can you take a week off any time you get ready? I'm just curious.

How often do you see Mary Reed? She's a good kid but I don't think she agrees with some of my ways. She got the idea somewhere that I was leading her Bill astray when in reality he was behind shoving me toward hell. I couldn't do anything wrong – just ask my mother.

Anyway I wish this Reed woman would leave me out of her gossip column. I imagine she gets all her misinformation from Mrs. Bill Williams and Darlene gets hers from a very unreliable source. There's a conspiracy against me somewhere and I think I know where it originates.

Let's you and I label this little discussion 'Confidential' until we can make an investigation.

I'd better write my Mama tonight. It's almost midnite so this will be all for now. Don't forget to write. Yours, Joe T.

May 29

Dear Janie. Ordinarily I don't wait three days to answer your letters but I've been more or less rushed for a few days what with one thing and another.

Bill has decided that we're going to Wun Wing Lo's (so help me, Mary, that's his name) Chinese restaurant in the village for some chicken. His food is about as good as anything our own mess hall mixes up so it doesn't make much difference to me other than that I'd better write you this afternoon for Wun Wing Lo's food doesn't always leave one in a mood to write pleasant letters.

A good friend of mine – one Steve Pound – I've probably mentioned him in some of my letters – has been overdue for four days now. And this is awfully damp weather to be wandering around out in the hills. Search and Rescue finally spotted a wrecked airplane and three opened parachutes nearby so it's a fair bet that all the crew bailed out O.K. I hope old Steve gets back O.K. for he's a pretty nice guy and in addition to that he owes me a hundred bucks.

When you mentioned that you opened my letter in front of your mother I became very uneasy for I can't remember whether or not I enclosed any of those cute little Esquireish clippings in that particular letter. Mrs. Griffith, being a mother, might not understand and be led to believe that I'm not a sweet innocent little boy. Which I am, of course.

So please advise so I'll know whether or not to wear my bullet proof pants when I come 'round to call on you when I return to the Promised Land.

I call it the Promised Land for they promised me when I left that it would be a long time before I landed there again. I'm beginning to believe they weren't foolin'.

It's been raining in earnest for a week or so now. I'm even afraid to get drunk any more lest I fall in a drainage ditch and drown for they're all full of muddy water.

We have some slit trenches scattered around the area to sleep in just in case the Japanese get any ideas about dropping a few sticks of dynamite around here and there.

The Flight Surgeon told me one morning while he was patching me up that we had more injuries on the field as a result of certain people (not mentioning any names of course) who get pie eyed and stumble into these trenches than we could possibly have were we bombed every nite. Which we aren't for our honorable enemy hardly has a leg left to stand on in Burma.

Finally had a letter from Gertrude. It seems she is suffering acutely from the man shortage. It's a damn shame I have to be bottled up in this hellforsaken place at a time like this when I could be keeping at least fifteen or twenty of these lonely females happy.

No, Mary, there are no French babes over here, damit. The only females I've seen other than a few nurses were about the color of a frying pan and altogether most untastely looking little dishes.

No, you needn't explain the design of your new dress – you just be in it and I'll be happy.

Bill is pestering me to hurry up so we can shove off. Have more trouble with that guy.

Be good and write soon and I'll be seein' you. Yours,
Joe T.

May 30

Dear Snookums. Just received your letter dated May 20 and decided to pass back eccentricies (sic) before the C.Q. comes over with his little flight schedule and flushes me out of my comfortable little den. (Completely furnished with three pinups of La Lana Turner and one Varga girl.)

C.Q. is 'charge of quarters', a duty in which an armed forces service member guards the front entrance to the living quarters—in this case an array of shelters known as *bashas*.

> So in view of the fact that I'm going to have to buzz into the blue with the iron bird this may be a short note but you can see I'm tryin'.
>
> Life is so cruel to Hump jockeys and their women and people like that there.
>
> Hey what brand of vitamin pills have you been using? It must [be] some powerful new formula to inspire even you to slave over a hot washboard for three hours! By the way – can you sew buttons on shirts too?
>
> You know I don't think you realize what a connoisseur of dress design I really am. Now for me to really do justice to the designer a pretty pink and white satin hanger would never never do. It will be most necessary for you to be in the dress. Pretty pink and white young ladies achieve a much better effect than an ordinary pink and white hanger.
>
> This add [presumably additional] request of yours for a picture of me is going to be very difficult for me to carry out but I'll see what I can do. With this in mind I made a couple of inquiries and learned that there are a couple of obscure little photo wallas [wallahs] in the village. <– (37 Indians, 17 dogs, 9 Chinese and 68 Sacred Cows) So I'll see what I can do but remember what with the inferior photographer and the inferior material he has to work with (Shuddup, I'm just being modest) the result will probably hardly be anything to show your little friends. We'll see.
>
> Did I mention that it's raining? Of course you could probably have guessed that for you can put your finger on practically any date on the calendar and give odds of 5 to 1 that it will be raining around here.
>
> For lack of something better to do I've been reading an exciting story this morning about a guy named Hopalong Cassidy whom I strongly suspect is no one but William Boyd in disguise. I've been very worried about whether or not that horse thief is going to get away but I don't have time to find out now for the C.Q. just got the drop on me. G'bye for now. Love and that sort of stuff, Joe T

For our grandchildren: William Boyd was an actor who played the character Hopalong Cassidy in a series of western films.

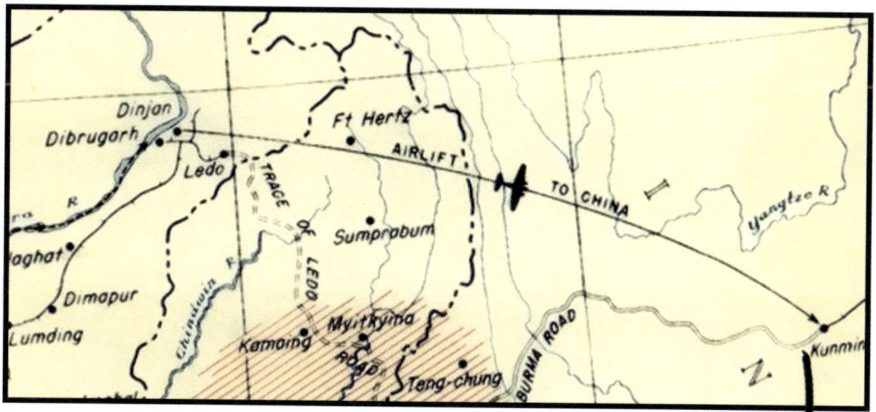

Magnification of the airlift to China (the Hump) in the right corner below

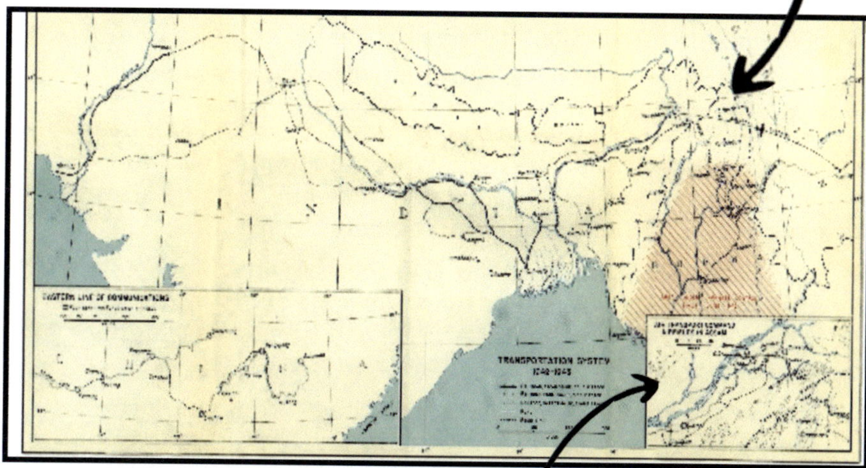

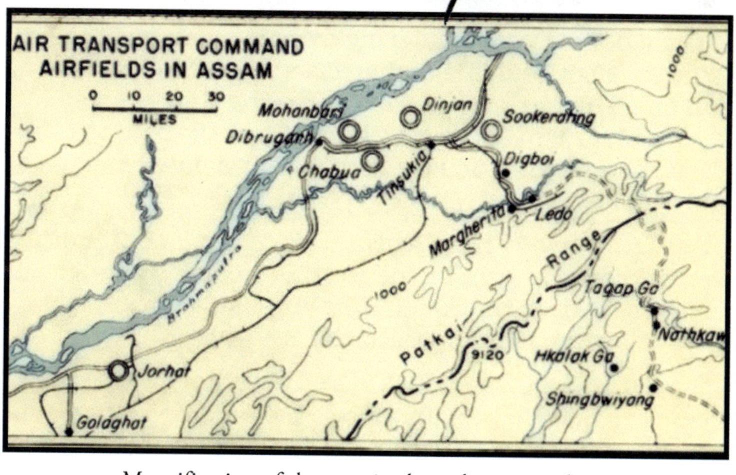

Magnification of the map in the right corner above

Anatomy of a Flight Over the Hump and Back

This letter is undated but was obviously written after Dad had been flying the Hump for a while. It seems to have been written to convey his Hump flying experience to Mom without alarming her unduly.

> India – It's two o'clock in the morning. Outside I can hear the drip-drip-drip of the monsoon rain slipping off the grass roof for the C.Q. has just flashed a light in my sleepy eyes with a "Wake up sir! Ship's ready!"
>
> Mentally I put a curse on the C.Q. for waking me at this unholy hour and, swinging over the edge of my bunk, snap on the light. My roommate groans and turns over, hiding his head under a pillow.
>
> I throw my stuff into my parachute bag – that extra pack of cigarettes, a candy bar, my heavy flying suit and my fleece lined boots for it's twenty below up there.
>
> Making doubly sure I have my oxygen mask for it's the difference between life and death at 20,000 – I hurriedly pull on a shirt and slacks and throwing the bag over my shoulder run across the street thru the rain to that glimmering light – the mess hall – for a quick cup of coffee and a sandwich.
>
> "Breakfast" over I find Brown waiting in the C.Q.'s office so we throw our stuff in the waiting weapons carrier and go jouncing over the rough mile and half to Operations, picking up our radio operator on the way.
>
> Operations is a blaze of lights and activity. Make out a clearance and flight plan. Check the none-too-promising weather, pick up a 'chute and we're on another truck for the dispersal area and the airplane.

There she is, big and shiny, glistening in the rain. Tonite we have a load of —— for China.

The engines purr over and check out O.K. so taking a tower clearance we taxi out on the runway and giving her 49 inches go roaring off into the night. She staggers into the air for this is the biggest two-engine cargo plane in the world and she's heavily loaded.

At some point in India Dad transitioned from the C-47 to the larger, faster C-46. The C-46 was a workhorse of the Hump airlift and was the aircraft type to which Dad referred here.

We climb up thru the scud and breaking out on top set her on course for China at our assigned altitude.

Just past the Love Xray marker beacon we run into some heavy weather. The ship tosses and groans and we start picking up ice. I can see it forming around the edge of the windshield and on the windshield wipers. And I know we have a heavy load on the wings for the airspeed is dropping off in spite of increased power. I peer out the window but I can't see the wings thru the blackness. Ice starts flying off the whirring props and pounds the sides of the airplane.

I turn to the radio operator and ask him to call the Traffic Control Station and ask them to clear us on top.

Traffic Control clears us 500 feet on top so giving her all the throttle she'll take we start a sluggish climb and finally break out of the clouds into the clear at 19,000.

The ice starts melting off the wings so we settle down on course at 19,500. I adjust the seat and relax for a moment for I'm bathed in sweat. I can feel it trickling down inside my oxygen mask and my nose begins to itch where the mask fits too tight.

Up here everything is calm and serene. There's a big yellow 'Texas' moon – big as a wash tub – and looking down at the silvery cloud tops slipping by they look not like clouds but like fleecy snow covered hills in the moonlight.

I glance at the engine instruments – everything seems to be O.K. – oil pressure, oil temperature, cylinder head temperature . . .

I can see the sky turning to gold off in the east over the nose. The sun is coming up.

The overcast is breaking up beneath us and as day breaks I can look down thru the holes in the clouds and see the rough Himalayas.

I tune in the Kunming beacon on the radio compass and we come riding in over the field in the clear. The overcast has dissipated with only a few scattered clouds polka-dotted here and there so we swing down to traffic altitude and come gliding in for a landing. Park the ship, grab a cigarette and an egg and a cup of hot coffee for the air is brisk – this field is 6,000 feet above sea level. Quite different from the sweaty climate of India.

Make out a clearance and with the plane unloaded and refueled get back in the blue for home. Back at the home base in India we check in with Operations after covering just over 1,000 miles of the toughest air route in the world. Pile back on the same weapons carrier and go jouncing back over the same rough mile and a half to quarters.

A leisurely meal, a cigarette and then check the mail.

Oh boy! I'm ratin'! a letter from Mary!

Listen to this, Bill. She says she's going to move into a tent in the middle of Main St. That's Mary! And – oh oh! What's this? Hmmm. She's been out with that Pirtle guy again. hmmm

After a shower and a cool shave I'll be back under my mosquito net getting a little sleep before that man comes around with that flashlight again. So Goodbye for now – Love, **Joe T.**

Dad once told me about an odd frequent occurrence at the China end of his trips. Chinese workers believed they were pursued by spirits, so they tried to run very close in front of the aircraft's turning propellers so the props would eliminate the spirits following behind. Every now and then one cut it too close, with unfortunate results.

Dad was downplaying the threat posed by weather. From *Skies of Thunder* by Caroline Alexander, pp. 92-93: "The Hump route bored through the chaotic intersection of three major air systems: moist, warm, high-pressure masses spawned by the Bay of Bengal to the west and the South China Sea to the east, and frigid, dry lows that swept down from Siberia. This unique convergence created the seasonal thunderstorms, monsoons, and ice that were chronic along the Hump route. Additionally, fierce winds blew out of the heart of Asia across the mountains and high tableland of Tibet, while from the west a subtropical jet stream blew continuously at high altitude.

The collision of these torrents of wind with the unyielding barrier wall of the Himalayas created tumultuous roller coasters of wind-wave effects. . . Flying the route in spring, when the turbulence and storms were generally worse, one pilot was lifted eight thousand feet in one minute as he crossed a ridge. But a powerful downdraft over high terrain gives a pilot little room to survive, while a ragged, lurching wave can tear off an aircraft's wings."

Pilots referred to their route over the Himalayas as the "Aluminum Trail" because of all the crashed aircraft along the way. The Air Transport Command officially listed 594 aircraft lost over the Hump, but that figure is universally regarded as incomplete.

The Hump Airlift operation was a logistical marvel and the US Air Force called it the first and the largest strategic airlift before the Berlin airlift with over 650,000 tons of supplies and men flown over the Himalayan mountain range in over 42 months with a plane crossing the Hump every minute and twelve seconds at the peak of operation. Airfields such as Jorhat, Chabua, Dinjan, and Mohanbari in Assam became pivotal points where aircraft were loaded with supplies before embarking on the perilous journey across the Himalayas to China. Assam's proximity to the Burma front and its access to India's extensive rail and road networks made it an ideal base for these operations. "Every vehicle, every gallon of fuel, every weapon, every round of ammunition, every typewriter and every ream of paper which found its way to Free China for either the Chinese or the American forces during nearly three years of war was flown in by air from India," the official US Army Air Forces history of the war said.

With one of the worst wartime survival rates ever, the airlift was achieved at the cost of over 1600 airmen in over 700 plane crashes. "Every 340 tons delivered cost the life of a pilot," wrote historian Francis Pike in his book, *Hirohito's War: The Pacific War, 1941-1945.*

June 1945

June 2

Dear Woman. Confidentially, I don't know how a certain one people I know rates practically 90% of my laborious correspondence. Couldn't be the Spring weather because we don't have no stuff like that there in &#*!!?** <— (India).

Maybe it's because you're the only people I can think of anything intelligent to (no cracks!) say to.

You might possibly have received the p.j.'s by now. Ye gods I hope they didn't come while your mama was there! Anyway maybe you have the dubious distinction of being the only gal in the house who owns a pair of Chinese pajamas.

And by the way – I got thinking that blue might not go so well with brown eyes. Offhand I can't remember seeing you wearing much blue. (When I remember you I see red!) Yeah I know – that was a dirty remark. Lemme see where was I?

So if you don't like 'em send 'em back and specify which colors you prefer and I'll take 'em over and see what I can dig up.

I got stuck in that China place with engine trouble again last night. I don't particularly care for that section of the country.

Heard any good gossip and stuff recently? I'm all ears. And lets not have any remarks about the abnormal size of my ears either. Miss Griffith!

Steve hasn't shown up yet. I'm getting a bit worried now. He's been missing for over a week and I don't see how he could last seven days without liquor. [Steve never returned.]

If you'll remember that I haven't been to bed yet last night you'll understand why I'm going to close here. Goodnight Now. Love always, **Joe.**

June 3

Dear Janie. OK! OK! So I ain't no gentleman! Can I help it if I'm jus' a low brow? Anyway I thought it was funny. But if you'll keep speaking to me I promise to be a good lil' boy. So there too!

My mama told me never to promise no woman nothin' but you're an exception. Particularly since I had two (2) letters from you today written May 23 and 24 respectively.

Hey! Maybe we should have a picture more often – say once a week – if it's actually going to inspire you to comb your hair and wash your face! Why I remember when – but maybe I better shut up. Twelve thousand miles really isn't a very safe distance when the wrath of a woman is around.

And by the way. Today 3rd June 1945 A.D. – I dug up a clean shirt (that's a monthly event too, what with the laundry situation what it ain't) and went down town and hunted up a photographer. It's all your fault. He took one look at me and said something to his assistant in Hindustan which I strongly suspect was something on the order of, "By the great Allah, what does this guy want with a facsimile of that sunburned face?" How am I to explain to a guy like that about how foolish American women are?

I finally got some snapshots developed. That's what those little squarish things are that fell out when you opened the letter in case you haven't figured it out yet. Well, it could have been the camera you know. Those aerial views I took about sundown one day on the way to China.

By the way my roommate, known as Bill, will hereinafter be referred to as Junior, since he's such a little squirt.

He's been a 2nd Lt. longer than I've been in the Army. I think I'll promote him next month. I'm starting my third year in the Army a week from today. To be real frank and earnest with you I'm growing a bit bored with it all.

Lissen here woman. I got somethin' I wants to be discussin' wit ya. It sez here quote "Men may be dumb but they ain't that dumb." unquote. Now lissen here! Men ain't so – why don't I shuddup – you always win every argument. I give up.

You scared [the] hell out of me with that chatter about taking up housekeeping before I got to the second part of the paragraph you know. What do you call that kind of a grammatical error?

Let me know when you're going to move and I'll send you some Spam. It's delicious about the first 3,000 times you have to eat it. I got tired of it long ago.

You will be greatly shocked to hear that it aren't raining here tonite. (That guy is even being nice and not playing his blasted phonograph.) The only uncomfortable thing worth mentioning is that it's been hot as ~~hell~~ heck (pardon) all day. If I survive this summer season it's going to surprise me and three Flight Surgeons and two nurses. I always catch everything that's going around and some that ain't. There's plenty of bugs around here that ain't. You couldn't be sick because the Flight Surgeon never heard of what's wrong with you.

Junior is theoretically sleeping but he keeps lying there talking to me and in general disturbing the normal run (shuddup!) of my train of thought. Yeah, I know, mine is a slow freight.

You're probably wondering when I'm going to be quiet and go to bed and leave you alone but I may as well finish up this tablet since you have practically 90% of it anyway.

Yep it would be pretty hard on feet and stuff if I ever had to walk home. Steve is out there walking around somewhere. I hope he's walking.

Leah must be quite some gal. Them is my sentiments too — Poor Howard! All of which goes to show you that it don't pay to believe anything a woman tells you. I've been used for almost everything but a door mat. I'm expecting that any day now. Maybe I should go out and get 'welcome' painted on my shirt just in case. Women are such unpredictable creatures.

I've still got two pages left but I'll save them for another day so you can be prepared for a short letter next time. Be good and if you move into that house be careful and don't get ptomaine (or something) poisoning. Goodnite Now. Love Always, Joe T.

June 8

Dear Mary. I've just finished my third shower (cold) for the day so maybe I can keep my temperature below the dew point long enough to finish a letter. Now don't get excited and decide that I've made a radical change in my way of life and adopted cleanly habits – nothing so drastic as that! It's merely the only way I can figure out to keep from being fried like an egg. Imagine me praying for rain! Imagine me praying period. To put it mildly the climate ain't exactly ideal around here.

I sent Junior off to the Link Trainer for an hour so it's unusually peaceful and quiet so I can meditate on this letter from you today.

Now lissen here woman. I did not desert you after one and one half pages for a cold beer. Junior twisted my arm and made me go. So give him hell. Speaking of cold beers – Hmmm. But ours is all gone. We get a case per man about the 26th of the month so we dash over to the club and put it on ice. It's usually all gone by the 28th.

I suppose you're on your way home today. If you see anybody I know tell 'em hello for me. That shouldn't be much of a problem for I don't know many people anymore.

This new house of yours sounds O.K. Annie must be quite some gal if she can cook and keep house. Too bad she already has strings attached. . . . Hmmm. them kind of women is scarce.

well, heck, I've got to go. At least I'll be up where it's cool. See ya tomorrow. S' long now.

June 9 – afternoon

Hi woman – Ordinarily I don't write my letters on the installment plan but you know how rough things are sometimes. Ordinarily I just don't write many letters. I never thought I'd come to this – being a slave to a fountain pen I mean.

It's so hot that I woke up this afternoon after only ten hours sleep. (Take the star out of your window, Mother, your boy's in the ATC.) [The ATC was the Air Transport Command. Despite their high risks and heavy losses, the ATC was not considered a combat command.]

You never told me about this little incident with Marjory Brown. I would loved to have seen that. On second thought I'd rather see you do it now.

Gee Whiz, everybody seems to be getting married. You hadn't better start getting ideas along that line, Miss Griffith! — You're too young and besides I would be displeased for some reasons which we won't delve into here.

And speaking of matrimony. Since when have I given anyone any encouragement . . . recently?

This would be a lovely time of year to be home. All the birds and bees and flowers and cool breezes. Let's quit thinking about — it makes me unhappy.

I've just been browsing through a recent letter here in which you were elaborating on what a beautiful night it was what with the "moon and the soft silvery clouds and the stars sprinkled around." Hmmm. That does not sound like you, Miss Griffith! I always thought you were the cold hearted type. Come, come, Mary, what's his name? I was afraid something like this would happen when you got away where I couldn't keep my eye on you. A poor wandering hump jockey is licked from the start to the finish — which is what I'd better do with this letter since it's almost time for the evening hash and canned beans. G'bye now. Yours always, **Joe T.**

June 12

Dear Woman. I betcha you're home right now and keeping that poor Pirtle guy guessing again. He's far more intelligent than I ever was if he isn't guessing. Of course that wouldn't be saying much for his intelligence. I should be burned up about the way you kick him around since he's a man like me and therefore on the same side of the big battle but for some strange reason I'm not.

You know I don't think I'm going to like that guy very well for a couple of reasons. You're too young to understand so I won't explain.

For your information, Miss Griffith, I just turned down an invitation to go to the Club for a cold beer so I could write this — uh — letter. I don't know why I make all these here sacrifices.

The only place I can think of offhand that I like less than India is Texas. Him all same-same. (I've been here too long. I'm even adopting the language!)

I went over to the river yesterday afternoon for a swim and came back with the nicest red suntan. This is the most uncomfortable suntan I've ever experienced.

Some native out at the river got sore when we borrowed his canoe and paddled up the river. We talked him out of taking our scalps since we didn't hurt his canoe any.

They had some big battles around here a few years ago before the Americans moved in. The Nagas would come paddling down out of Burma and take home a whole boat load of heads to hang in front of the hut of the particular squaw they happened to have their eye on at the time. That's an awfully queer way to express affection.

The lights just went out so we had to dig up the kerosene lantern. It puts out about 1 1/2 candlepower and about 700 BTU's heat.

I shudder when I think that I have to stay here until December. Unless the Japs get wise and throw in the honorable towel. Then I'll probably be in this hole another year hauling all that stuff back I've hauled over. That would be the payoff. Ain't I got troubles, Mary?

Yep, them is awfully handy shoulders you have. I've got to get some sleep now. —Write — Luf Always, **Josef.**

June 14

Hello Sweet and Lovely. I don't know who flies the mail plane from Calcutta but he must be a good guy. Anyway he brought me two letters from you today. (Yesterday I was cussing him out.) Those were dated June 2nd and 4th. Ten days isn't bad at all. I got one last week dated October 2nd. It had been to England and France and God only knows where else. And, too, a Christmas package came from my sister yesterday, June 13, mailed in October. Yep, you guessed it. It <u>had</u> been food. I dumped the whole mess in the garbage can.

But down here it takes your letters from eight to twelve days. So they aren't exactly ancient history. Of course you never can tell what a <u>woman</u> might do in eight or twelve days. They're queer critters.

You might like to know that your mail goes to Miami to Bermuda to Casablanca to Cairo to Calcutta to here — wherever that is. You guess. (The &@!!&@ base censor, you know.)

I would like to have a job flying it from Dayton to Miami. But then if I was there you wouldn't need to write.

My letters are subject to spot censoring. The base censor pulls a couple out of every hundred or so and reads 'em. The old busy body. The time he whacked a piece out with his little hatchet was when I told you I was in the Assam Valley. If this censor would take his feet down off that desk long enough to read the censorship regulations he would find that it's perfectly O.K. to say Assam Valley as long as I don't say which end. Some people are the stupidest @&!&&!!

It's raining again and I'm happy for it isn't so cursed hot. It's so damp matches won't light and cigarettes will hardly burn if you can coax a match to light. Like I said the climate ain't ideal around here. It's always one grade better than worse.

I journeyed into town with Junior this afternoon and picked up that picture you've been wanting to hang on your garden gate. The darn guy stuck it on an old hunk of cardboard which wasn't according to my plans so you'll have to suffer with me. Maybe you can pull it off with hot water or something. Anyway I'll mail tomorrow if I'm not flying. You'll probably get it by Christmas at least.

A portrait photograph has come down to me with the cardboard backing still affixed. It shows a *very* young Dad in uniform with CBI insignia, wearing a pith helmet and holding a pipe. Mom apparently decided not to try to remove the cardboard.

I saw an old instrument instructor of mine over in China the other day. I never did particularly like the guy so I was happy to see him over here. That's the worst thing I can think of to happen to a man.

Before I left the States I was itchin' to get over here. Well here I am – just itchin'. But I think I pulled that one on you before. If I tell you these stories more than three times let me know and I'll try to dig up some new ones.

Offhand I can't think of any good looking young bachelors who need a private secretary but if you wouldn't be so choicy (sic) and merely say young bachelor period I could probably use – anyway when I get my school-for-the-mentally-unbalanced started after the war plus six maybe I can find a place for you. On second thought I'm sure I could so don't go and sign any long term contracts of any kind.

What's the news of the sergeant-being-used-as-a-doormat? Ho Hum – Women!! They sure are wonderful when there ain't any around for hundreds of miles though.

One thing I never could understand why a male Indian looks like a human being and the women don't look like anything in particular. They're alive and that's about all.

We had chicken and French fried potatoes at the Chinese restaurant tonight. At least it didn't come in tin cans and that's a change.

Did I ever tell you about this new invention I have for dehydrating water for use in Desert warfare? It's compact and easy to carry and all you have to do when you get thirsty is put it in a pan and add water. I think I'm going to have some trouble getting the formula patented – when I figure out the formula – but you know how they laughed at Edison.

Dad wearing a pith helmet and holding a pipe in uniform with CBI insignia.

Dehydrated eggs are a wonderful dish. They have the delicious flavor of a mixture of hay and old rubber tires.

I hardly know when Sunday comes because I usually work just the same. No, Mary, I ain't a atheist but it's an awfully lot of work to look at a calendar. The thing I don't like about Sundays is that they close the bar.

I'm going to send this letter to 313 Gramont and if you decide not to move there I'm going to be mad. And by the way when you look up the phone number you might send it so I can write it down in my little black book just in case. You never can tell where I might turn up next.

I don't think you'll ever have any cause to get pickled like a herring, Mary. I can't imagine you like that anyway but as I said you never can tell what a woman might do –.

This will be as good a place as any to say – Goodnight Topsy, Amor and more, Joe T. (over) This letter is so thick the base censor will open it sure in search of hidden maps, secret documents etc. And after all those unkind remarks we made about him, too! Tsk! tsk! I'll probably be hung no less. Or shot at sunrise.

June 23

Dear Brat. Hmmm. Just as I thought! The minute I turn you loose in the vicinity of that Pirtle guy I don't get a letter for practically weeks and weeks. My faith in women has been destroyed. Again. . . .

Of course someone might have got provoked (and I can easily see how that could be) and busted your arm. I keep telling myself. I'd probably be better off if I shut up and quit telling me these things.

This is a particularly depressing Saturday evening.

Junior and I went to town for dinner tonight so it started raining like hell, naturally. We got good and wet coming home in the jeep. That's always a pleasant way to start off a Saturday evening. Ho hum.

Bob is trying to entice me over to the club for a beer. Junior is in bed presumably sleeping since he has to fly at midnight but he's doing more talking than sleeping.

So if this letter sounds a bit more disconnected than usual just remember all the side issues I have to put up with.

By this time you should be nicely established in your new house. Well, how do you like housekeeping? Or am I likely to get my arm broke for asking.

Just be sure to use Ivory soap so your hands don't get all rough and red you know – and while I'm thinking about – How about getting out your little hatchet and trimming those finger nails you used to threaten me with sometime before I come home. It gives me such an uneasy feeling to face a lady on the field of battle armed with those gadgets.

Don't you know it's unlawful to carry a weapon in Ohio?

I was just thinking tho — maybe you'd better save 'em to do this little post war perteckting business we have in mind.

Just as long as you don't change your mind and turn 'em on me. A poor man never can tell what a woman might do. Particularly a woman I know whose initials are M.G.

I just remembered I've been owing friend Mary Reed a letter for a few dozen days now. I really must write her some week soon to find out what I've been doing lately.

Have 36 Hump trips now. I'm getting bored more than somewhat with this miserable place. If we ever run out of liquor it will be a dull life indeed. [   ] <— stifled yawn. The only interesting thing about this place is that it isn't likely to be.

I'd better run along now and I leave you to slave over that hot stove. Goodnight you all. (I'se from de suth'in pa't of Indiana you all know honey chile.) Yours, **Joe.**

June 26

Dear Mary. I'll take it all back. I'm humbled. I'm sorry. I'm apologizin'. You really did write me a letter while you were home.

Just between you and me I'd never expect anyone to spend valuable vacation time writing letters but then you're no ordinary person in at least a couple dozen ways I could mention.

Anyway I had to wrack you back for something and couldn't think of any better reason since, as far as I know, you've been a good little girl for the past couple years. You might even grow up to be a lady yet —. Ouch! Get those fingernails out of my neck, Miss Griffith! Ho hum. These feminine tempers. tsk tsk! But oboy do I miss 'em! Life jus' ain't the same.

I've been acting as collaborator and trying to make peace this afternoon. Junior has adopted a belligerent attitude toward Bob because he thinks Bob has been drinking all of Junior's beer while Junior is out on a trip. Complicated aren't it? And shut up Mary, I have not been snitching Junior's beer.

Bob had to go down to see the Provost Marshal this morning. It seems he got pretty well oiled up over at the club last night and decided he didn't like the looks of some Captain who was sitting at the bar so he tossed him out on his ear. The Captain took a rather dim view of the whole affair and called up the Provost Marshal – thus the reprimand. Yep, my friends are remarkable characters.

Don't be troubled about friend Doug. He's mad at me too, and hasn't answered my last letter of a year ago. My mother has seen him once or twice recently and remarked in a letter today that he's the most disgruntled boy she's ever seen. It appears he's looking at the world thru dark brown glasses. He'll snap out of it when he grows up.

If you've been fired when you get back to Dayton you can start working for me. However all the correspondence must be directed and addressed to me. You'll find the address on the front of the envelope.

Why Mary! How come you didn't tell me about your engagement to this guy in Dayton. After all I would have been somewhat interested you know. Now you know where those supposed ventures of mine originated. tsk tsk.

Who did Seymour Hunter marry? I think someone told me once but I've forgotten. I probably wouldn't remember her anyway. I never write to anyone anymore but you and Gertrude and once in a while Mary Reed when I want to find out something about Bill. Poor guy. She wouldn't even let him go up town with me alone when we were home on leave last summer.

I must be quite an example of look-what-will-happen-to-you-Junior-if-you-don't-be-a-good-boy-and-eat-your-spinach-you-might-grow-up-to-be-like-that-guy around Sullivan. Frankly it just worries me to death. tsk tsk.

When I tell you that it's nearly ten P.M. and I haven't had any sleep since 7 AM <u>yesterday</u> morning (In other words, I haven't been to bed yet last night) You'll understand why I'm going to say Goodnight. I haven't had my shoes off for 36 hours. Me and Errol Flynn. Or was it General Custer. I went on a trip last night and didn't get back till this afternoon. My feet are killin' me. They get <u>so</u> tired propped up on the instrument panel while the automatic pilot flies. Ho hum.

Well Goodnight now. And the next time you go and get engaged to some guy I want to be informed. Understand?
Love Always, Joe T.

P.S. After long consideration I finally decided to do the gals you live with a great injustice and mailed that picture you've been pestering me for yesterday. What I want to know is – when are you going to give with yours?

The Death of a Friend

This letter was written long after the war, but it concerns events in early 1945 so I have included it here. It is an intriguing question why Mom and Dad still had it to pass on to me. Apparently, it was never sent. Were the addressees no longer living or no longer at the address Dad knew? I don't know and I doubt I'll find out. Although it is about a sad subject, it's valuable to me for the information it includes.

A short summary of the background is that Dad and one Steve Pound met in training. They shared a relatively uncommon family name and became friends. In the course of their training, they went their separate ways but found themselves together again in India. Dad survived but Steve did not. Dad's address book includes the entry: Steven B. Pound, Jr., 1456 Ingleside Ave., Jacksonville, Fla. There is a second entry (different handwriting): Stephen Baldwin Pound Jr., Rt. B, Live Oak, Fla.

Although Dad here recalls Steve's loss as happening in March or April, we know from his May 29 and 2 June letter that Steve was lost in May 1945.

Linton, Indiana

October 30, 1950

Dear Mr. & Mrs. Pound, Your son Steve may have spoken of me in his letters to you for we were good friends in early 1945. After more than five years I would like to tell you something of our last days together. I want to give you what little information I can and in turn learn from you what you have heard from the War Department concerning the circumstances of Steve's death. I'm sure you will be

Steve Pound
Dad's friend who was killed

interested in anything I can tell you but hope I do not cause you to feel too badly in writing.

I first knew Steve at the San Antonio Aviation Cadet Center where we slept side by side in the same barracks for about a month in the fall of 1943. We went from there to Preflight and then on thru different flying schools. I went to Europe in October 1944 and stayed until January and then on to India.

I arrived at Dum Dum Air Base just north of Calcutta about the middle of January 1945 and about the first person I bumped into was Steve Pound. We moved into a tent together. The tents were a British make with a double top to insulate against the heat and with concrete floors. It was not at all an uncomfortable way to live in that climate. We had a fine mess hall equipped with an excellent cook. We also had a Hindu bearer boy named Abdul Hamid who shined the shoes, made the beds, took the shirts to the laundry etc. We treated him as one of the family and he shed genuine tears when we left.

On days off we went swimming at a private pool called the Calcutta Swimming Club and played pool at the Great Western Hotel. On several occasions we had steaks at Fir's Restaurant, a world famous eatery.

Around the first of March we were both transferred to an airbase in northeastern India (called Mohanbari) to fly the Hump. Mohanbari was on the banks of the Brahmaputra river and was about midway between Jorhat and Sodiya, which two places generally appear on any good atlas map of India.

We were introduced to the C-46 here and started crossing the Hump to such places Kunming, Chengkung, Chonyi, and Luliang, all airbases in western China.

At Mohanbari we moved into adjoining huts (called bashas). Both bashas appear in the snapshot I'm enclosing. Steve lived in the one immediately behind the native grass cutter and I lived in the far hut across the path. These huts were again well suited to the climate and were fairly comfortable and cool. Many people on this base thought we were brothers. [The mentioned snapshot and other photos were with this letter.]

We had not been at Mohanbari long [when Steve] went on a trip to Myitina (spelling approximate — pronounced Mitch'in-aw) in Burma as copilot on a C-46. ["Myitkyina" is correct and is used hereafter for clarity.] (Myitkyina was the northernmost point of penetration of the Japanese in Burma but by this time they had been pushed far south towards Mandalay.) The number of the airplane was, I believe, 796. They left Myitkyina on the return trip to Mohanbari, a flight of about 2 hours, and were never heard from again.

I went to see the Intelligence Officer of the 1352nd Search and Rescue Squadron the day before I left to return to the States in November 1945 and at that time the wreckage of 796 had not been found although they were searching constantly for wrecks on every day the weather permitted.

On this particular day (which I believe was sometime in March or April 1945) the weather was bad but that was no particular deterrent because about 40% of our flying was on instruments. [We know from Dad's letter of 2 June that it was actually in May.] I have no doubt, however, that the weather contributed to the accident. The nature of the country between Myitkyina and Mohanbari is very difficult to imagine, some of it so thick with vegetation it would take a ground party a week to travel a few miles. The policy of the search and rescue squadron was to spot a wreck from the air and then send in a ground party. I believe everything was done that was possible.

When 796 was first missing I knew that Steve had about a 50-50 chance for about half our crews were able to bail out, walk to the Burma Road and catch a ride home on a truck. After a month had passed and none of the crew showed up I knew the chances were rapidly diminishing. I could not bring myself to write you at that time for I had no truthful encouraging thing to say.

Steve talked often of his sisters, Joyce & Sara, and was always either buying something to send them or thinking of buying something to send them.

I'm enclosing a snapshot of Steve which was taken at the Reno [Nevada] AAB in the fall of 1944. This is the only one I have. If you have one like it I would like to have this one back. I'm also enclosing a shot taken on top [of] the overcast on the Hump and one of the main gate at Mohanbari." [The letter stops here; there is no closing. It seems never to have been sent.]

Dad never spoke of Steve. I suspect this may have been one of his most painful memories of the war. In her 1986 recollection, Mom offered the following:

I'm afraid I have only hazy memories about Steve Pound. I know that when Joe went off to WWII he did not expect to return and that was all right. He was the epitome of the young idealistic warrior. He wanted more than anything to fly a hot fighter plane, but this was not to be. By the time he finished flying school, planes were needed to fly supplies into Europe and to fly the wounded out. This is what he did when he was in England and France.

When he was in India, he formed a close friendship with Steve Pound and when Steve was lost, Joe hoped for a long time he would walk out to the Ledo Road and find his way back. When this did not happen, I think he for some reason felt guilty about it. In some way I believe he felt, either consciously or not, that Steve had died in his place. As I said, he had not expected to live through the war and was somewhat surprised when it ended and he was still here. He had always intended to contact Steve's family, but to my knowledge did not. This may explain the letter you found — he might have written it with the intention of sending it but did not because somehow it never seemed right to him that Steve was dead and he was alive. This is speculation but the best I can do.

July 1945

July 5

Hi Woman. I don't understand how I slipped up and let you get two letters up on me but it's unusually cool and comfortable this afternoon so maybe I can catch up.

I set some sort of precedent last night and went to bed for the first time in two days. Slept fourteen hours and woke up in a <u>very</u> nasty mood. but after reading over your letter-that-came-this-noon a couple of times I'm gradually getting back to abnormal. By the time I get this letter written I'll be my old self again.

I borrowed this paper from a friend who comes from Texas. He calls me an uneducated bum because I didn't recognize a Bob Wills and his gang record. Texans are such queer people.

Speaking of records. Someone next door is playing "Stormy Weather." Yep, it sure is rainin' all the time. Reminds me of something I read somewhere (Sure I can read) that said quote, Into each life a little rain must fall, unquote. Well I've had about 30 inches in my life already this season which is enough to last me for the rest of the war plus six months.

And by the way. when I speak of Indian women I mean the India kind of Indians. No offense meant [to] your Cherokee ancestors. It appears to me that the Cherokees have turned out some remarkable young ladies. I attribute my questionable hereditary character to a long line of Irish farmers and an aristocratic group of English chimney sweeps.

Like many families of the period, my mother's family lore included a putative female Cherokee ancestor. Through DNA testing I determined the lore was false, but that research was far in the future from when these letters were written.

July 6, 1945

Hi you again. I had to go to work yesterday – No it definitely was <u>not</u> my idea! I don't mind working when I have some work I like to do but I've never been able to find any of that yet. Ho hum.

It came today – the picture. It's very lovely – the dress I mean. And the model – I don't want her to get conceited but I've been thinking – how can anyone get so pretty in just two years? (Please enclose 10 cents in your next letter.) Seriously though, Mary, it's a very nice picture and thanks again for sending it. You've still got that there old gleam in your eye, tho, that always warned me to run and hide behind the nearest brick wall.

And by the way – how about not getting your hair cut before I get home. I like it this way. It will be so nice to pull when you start to unsheath those wicked finger nails.

This is a nice folder, too. I'm very happy about the whole deal.

Now lemme see what intelligent questions (that will be 10 cents more, please) you asked in your last letter.

Hey! You aren't deserting me for the ranks of the enemy are you? At one point here you pause to quote, "chuckle in fiendish glee," unquote, at my deplorable plight involving a certain Illinois woman. (I'm wondering if the construction of that last sentence would stand up under the unmerciful eye of the honorable Miss Egbert.) Them wuz the daze. Verbs, pronouns, and adjectives and whistling at girls on the stairs.

But to get back to the original question – what goes? Just when I thought you and your nice fingernails were on my side too. It all goes to back up my ancient old theory that the female of the species is undependable in a crisis!

Maybe I can blackmail you. But no that wouldn't be gentlemanly. I'll just sign a pact (the influence of the San Francisco conference on me) with you. If you won't run out on me I'll try and help you convince this Pirtle guy that perhaps he should go get interested in someone else little girl. I would enjoy that.

But then you probably don't need any help. You never seemed to have much trouble convincing the poor hapless males who got tangled up with you that black was white.

I'll have to dig up some other angle to get you back on my side.

For your information we had some of that delicious Spam for lunch today. My diet consists mostly of Spam and atabrine tablets and Spam and canned tomatoes and Spam and canned beans and Spam etc. (This could go on for cans and cans.)

So this fellow Carl starts all your letters off to me? You know I think it would be a good idea to work him a little harder. Yeah, I know – I'm never happy.

I see your point about the p.g.'s [?] what with it marked on the box and everything. There is an Army regulation to the effect that all packages must have the contents listed on the cover. The object of that regulation is this: Now if you and I happen to be Japanese agents and I desire to send you some Maps and Diagrams and Fortifications of our tight little airfield I must plainly mark it on the cover: MAPS, DIAGRAMS, AND FORTIFICATIONS ENCLOSED. Regulations say so! Then without going to the trouble of opening all the packages the base censor can note the contents of my package and remark, "Hmm. Spies!" And promptly have you and I shot at sunrise. It's all very ethical.

It says here that men are the most undependable, inconsiderate, ignorant, good for nothing etc. etc. If it wasn't so late I would go over in the corner and sulk for a while for this injustice. I haven't seen a white woman in such a long time I can't recall any of their unpleasant characteristics so I can't retaliate in that manner. Don't you think you are sort of taking advantage of me? But then you are a merciless female so I can't appeal to you on that score.

Had a Christmas card this week from Mary Reed. The mail service is remarkable. But I think I've griped about that before.

I'm going to be finished here and ready to come home in about three more months so I'll save a lot of time and ink and tell you all these other things I have to say when I see you – say in November. So be good and write and thanks again for the picture. Love always, **Joe.**

P.S. I'll note your phone number in my little black book under "Hobbies." You never can tell when I might be in Dayton with time on my hands.

July 7

Dear Mary. Oh, so now I'm a varmint? Oh well, I've been referred to as just about everything from rats on down by various numbers of my friends so the general term can't hurt much.

And don't you know I'm famous for ruining the white and shining reputations of just literally dozens of sweet innocent little girls? I'm what you might call a heel. And be quiet you. You don't have to agree with me. I'm just saying those things because I don't want people to find out I'm a gentleman. They would be disappointed.

Hello.

I ran out on you a couple of hours ago with all good intentions of making a trip tonight but the mechanics have my airplane scattered over about two square acres so while I'm waiting I may as well pester you for another dozen or so pages.

Fifty years the girl says and perhaps she'll be a passable cook! Just when I was drooling at the thought of a home cooked dinner thrown together with your own two lily white hands! My stomach is a bit temperamental so maybe I'd better not chance it.

By the way did I tell you that friend Junior is in the hospital with a fractured thumb and a broken jaw? Poor Junior. No, I didn't hit him with a brick for amusement. He didn't even get oiled and fall in a slit trench. He was playing baseball one afternoon and ran into another guy trying to catch a fly ball. Now he's living on a diet of soup and water – thru a straw. Poor Junior.

Steve was killed. He crashed and burned. The Search and Rescue Squadron is so far behind now you have to wait your turn to be rescued when you go down.

This timing does not square with Dad's 1950 letter, but it occurs to me that in 1950 he may have wanted to spare Steve's family the grisly details or, on the other hand, this letter may reflect incomplete information.

Now let's take care of the Pirtle Situation. Now if I were you I would – Lemme see, now, just what would I do? I seem to recall that I've been faced with a problem of this nature for several months now. If you figure out a workable formula send it to me quick! Ain't it awful to have that hunted feeling? I'd much rather have that hunting feeling. It's much more interesting.

I can't understand why she keeps griping because I don't write. She should have received that letter I wrote last month by now. Ho hum.

I've been very hard at work being lazy today. I busied myself planting some flowers around the Basha and trying to discourage a family of ants who decided to requisition a corner of our porch. Yep, this is a rough war I'm fightin'. Fifty years hence when my little grandson (I'm being optimistic now) climbs up on my withered old knee to play with my long white beard and listen to my watch tick and says, "Grandpa, what did you do in the last great War?" I'll probably strangle the little brat then and there if my gnarled old hands aren't too ravaged by rheumatism, malaria, and malnutrition.

And while we're speaking of Malaria did I ever tell you of the Great Atabrine pills?

Since our honorable enemy, the Jap, has practically all the world's quinine sewed up the Medical department had to figure out something new for a change – hence Atabrine. Theoretically it prevents the symptoms of malaria such as chills a[nd] fever so you can go right on working and never realize you're sick. Anyway we have to take one a day. They are a foul tasting little yellow pill and after one takes them daily for a few months the skin starts to take on the color of the pill – a very bright yellow color. It's getting so that when I go to China now all the Slopeys accept me as just one of the boys. Just call me Poo Fuey.

An October Christmas package came today from my mother. It was a little tired and travel weary.

Your picture made good time. Almost as good as a letter.

The Fourth of July passed and I didn't even have any firecrackers. I was disappointed.

I'll run along now. Consider yourself sufficiently pestered for one day. If you should happen to find an old scrap of paper lying around I'll be happy to receive it with some writing on it anytime. Your ever lovin' Joe.

9 July, 4,000 ft. over India – 6:30 P.M.

Dear Mary - Yep, woman, I'm all up in the air over you.

4,210 feet to be specific. Ordinarily I choose a more appropriate place to write my letters to you – say like locked in a dark closet at midnite and writing by candlelight.

I'm flying for the engineering dept. all this week. Putting slow time on new engines to break them in before they're sent over the Hump and stuff like that there. I took the job to rest from the Hump but think I have to start flying the Hump again to rest from this. I've been in the air six hours already today and still have 2 more to go. George is flying. George is the automatic pilot. He's the hardest working guy in the Air Corps.

Have some pretty good music on the radio. You know – that soft and mellow stuff you insist you're susceptible to just like an ordinary human being. You'll have to show me. I'm from St. Joe.

I would answer some of the questions in your letter but I haven't received it yet, darn it. Are you short of stamps again?

I've been up and down this damned river so many times today I'm getting tired looking at it. Ho hum. Did I mention the fact that I'm fighting a rough war?

How is Gertrude getting along in the painful process of getting educated and trying to catch a man simultaneously? Just to show you how much I miss her – I've even forgotten how to whistle in that old way. Perhaps it's just as well – Now that I'm growing up someone might take me seriously and bust my arm.

Have you learned to cook yet? Maybe you could send our new mess sergeant some pointers. I think he used to be a professional criminal – he certainly murders good food. O.K., O.K.! So it ain't funny. Nya nya!

I'll finish this when I get my feet back on India. Goodbye for now.

15 July

Dear Mary. There's no need for me to go to bed for they're going to be here after me to fly in a couple of hours. That's the best excuse I can think of offhand for pestering you for a few pages. Aren't you lucky? And don't answer that – I might be the sensitive type.

I haven't been over the Hump for a week so I'll probably get lost. Most of these battered old flying machines have so many Hump trips all you have to do is climb them up to 16,000 and they know the way from there. We have an airplane landing in China every five (5) minutes.

I think our mess sergeant must be sick. Poor boy. He slipped up and served us some canned turkey this evening. Where else could I eat canned turkey and buy fresh pineapple and coconuts for 8 cents each and live in a picturesque thatched hut along with 2 pet lizards and 7,276 ants? I've never had it so good. Ah, India! (Someone run and bring the Doc quick before I start talking to myself!)

Bob's on a trip, Junior is in the hospital, and I'm lonesome. On top of that they're remodeling the bar (it was in a bad way after Bob's last binge) and I can't even go over and gaze lovingly into the amber eyes of a frosty brew and feel sorry for me!

Everyone else in the area has gone to the movie. I got bored and went this afternoon.

Bob needs 12 more trips to come home. He's got 'em marked in big black figures on the wall and scratches one off every time he comes back. I sit here and watch the list grow smaller and turn all green around the ears with envy.

He should be back by now. He's been gone over twelve hours. I hope he didn't do anything foolish like jumping out in the hills. What with the physical condition he's in I don't think he'd live thru the first two miles walking.

When is Wayne [mom's brother] coming home? Those poor guys are going to have it rough when they're forced to go back to that old United States. No cigarettes, no good C ration meat and vegetable stew no nuthin'! Ho hum.

I believe I'd even be happy in Texas. That's how low I've fallen. I'm not at the bottom of the barrel – I'm under it. You'd better write me a seventeen page letter quick so the mail clerk can dig me out to deliver it. On second thought make it eighteen pages. Or twenty.

More guys have been around to admire your picture. I keep trying to explain the design of the dress but most of them are uneducated rascals and don't seem to appreciate the beauties of dress designing at all. They just keep muttering, "Who cares about the dress?" I'm getting insanely jealous so I think I'll hide it under my mattress or some such place as that.

The lights just went out for a few minutes. They come and

go like a Coca Cola sign. They went out for two hours one night just when I had half my face shaved. That was the night I lost my left ear.

I'm getting so wild I don't even believe this stuff myself now so I'd better stick this in an envelope and send it on its merry way before I start telling you about the time my Uncle Petey's sheepskin coat had twin lambs. One white and one black. Only the black one lived. Baaa-a-a-a-a. Goodnite, Topsy. Love etc. **Joe.**

P.S. And before I forget it, Miss Griffith, when are you finally going to bust down and write me a letter? (This was followed by a stern, unforgiving look at your picture. But you just smiled and looked wise.) I can't even make any progress arguing with your picture. I give up. Goodnight.

19 July, 7 P.M.

Hi Sweetstuff. I've cultivated and watered all the flowers along the walk since chow. I've cut all the grass in the back and I've showered and shaved and finished the last can of beer I can unearth and in spite of the fact that I was out on a trip last night and didn't bother to go to bed today I still feel wonderful. (Just a few symptoms of old age – aching back, sore feet, etc.) So for lack of anything more sensible to do I've decided to surrender to the inevitable and plunge into one of my literary escapades of which you are the poor luckless recipient.

All the generators have gone haywire again so I'm laboriously scrawling along by the light of a flickering old kerosene lantern. So if I break off in the middle of a sentence think nothing of it – I've merely wandered off the edge of the paper in the dimness.

I opened this letter of July 7 and bravely waded into paragraph one. Stopped, choked, gulped hard, and started over after the second time through I drop the letter and dash out and dive in the rain barrel in back. Bob runs out and drags me out by the heels and suggests that perhaps I should come back and read paragraph II. Then I discover it's all a farce cooked up over an ironing board by friend Leah. At this point I make another break for the rain barrel but Bob trips me at the door and I'm foiled again.

By the way, Leah, have you had your neck broken yet this season? Ho hum. Women!!! Always conjuring new ways to torture some poor hapless male. Have you picked up a new doormat for your house yet? Tsk tsk!

With luck I'll be home in November. I can't remember ever being very lucky though. Anyone who manages to get tangled up with a shipping ticket to India is one of the world's most unfortunate creatures. It shouldn't happen to a dog. And be quiet — I am not a dog! I see here in this famous letter of July 7 where I've been referred to as a cheap skate. Hmmm. Miss Griffith, how many times do I have to tell you — flattery won't get you no place!

Next Day

Hello. I wandered off to bed last nite and stayed for twelve hours. Now don't get sore pal, it isn't that I love my nice hot uncomfortable bamboo bed more than you, it's just that I couldn't keep the one good eye I have left open wide enough to see what I was writing and was afraid this old pen might slip up and say something I didn't want you to know while I wasn't looking. I have a lot of deep, dark, mystic secrets you know. I'm about as opaque as a summer day to the discerning eye of a wily female.

What do you mean, young lady, when you insinuate that I'm partial to loafing? Why I — lemme see now — anyway by the time I write you the next letter I'll remember when I did some work once and wave it in your face. Come to think of it, it is awfully nice to sit around in the sun and dream up all the ambitious things you're going to do next week.

My premeditations are on a little longer scale though and extend on through next week to next November. The first thing I'm going to do is run out and latch on to a big quart of ice cold milk when I hit Miami. I haven't seen fresh milk since last October. (Excuse me while I go into a trance and mentally take a bath in a big lake of the stuff.)

Next I'm going to rush out and grab the first pretty girl I see and kiss her soundly. Of course it won't be my fault if I don't happen to bump into any pretty girls before I get to Dayton. (Excuse me again while I go into another trance.) And stop hitting me in the face with that daid mackeral! (sic) I ain't calling on the speerit (sic) of my great grandmammy!

You should have received that gruesome caricature I put in the mails by now. Perhaps that's why I'm not getting any mail. You're probably still convalescing.

I'm going to have to make another trip tonite damit. Just once I'd like to see what it's like to fly by daylight around here. I'm suffering from a malady called "fluorescent eyes" caused by long hours of concentrating on rows of fluorescent lighted instruments. My eyes glow in the darkness like a neon sign.

And it is not just my wolfish nature! Remember a gentleman is just a wolf with patience anyway. Them is the guys you gotta watch. Enof of this prattle. Write darn ya. Love always, Joe T.

24 July

Dear Mary. I don't know why I slave away over this hot pen day after day singing you my blues – particularly when I can't seem to persuade you to reciprocate in kind.

I see here where I've been unjustly accused of laxity in writing. It appears to me that this makes no less than three (3) letters in the past five days and I ask you – what do I get when I stand up and face the miserable mail clerk? Nothin', that's what! And in your own words, quote, "It's revolting!" unquote.

If something doesn't break pretty soon I'm going to begin to suspect you've deserted and joined the ranks of the enemy. Both of 'em.

Sure is tough for Gertrude to be penned up in that awful community in the vicinity of Sullivan, Indiana, U.S.A. I don't know but I believe I could bear up under it for a few weeks at least, don't you?

You've also done me a great injustice (No. II) by insinuating (maliciously) that an increased beer ration was the cause of the emptiness of your mailbox. Miss Griffith do you honestly believe that I would allow a paltry case of beer to come between two such tried and trusted pals as you and I? Just when I thought you alone had faith in me.

For your information, by a clever bit of ingenious promoting one dark night this week Bob and I managed to lay hands on no less than five cases of mellow Budweiser. Again I bring your attention to this third letter of five days in spite of the luscious beer situation.

Excuse me while I go off in the corner and feel sorry for me. I can't win against a designing female. Ho hum. War is so cruel.

It's getting too dark to see out here and it's too damned hot to go inside and turn on the lights so I'd better run along. Write darn it. With Love, Joe.

25 July

Dear Mary. I sit here drooling in defiance as I gaze fondly at this full quart of mellow Bond and Lillard bonded bourbon and chuckle in fiendish glee when I think how I'm disproving your foolish theory that I'm more addicted to alcohol than to pretty young brunettes with brown eyes. And in particular one whose name is M.J.G.

And now that I've proved your conclusions on that subject erroneous let's go on to something more unpleasant – such as the weather. It's hot and sticky and uncomfortable and disgusting and I'm in a very belligerent mood as a result so don't give any argument please, Miss Griffith, for I would probably break your arm at the slightest provocation. Of course the fact that I'm some 12,000 miles away might cramp any act of violence on my part but stop smirking anyway!

28 July P.M.

Hi You. All I got to say is – it's about time you busted down and wrote a letter! It got here yesterday. The mail clerk was so happy. He was getting tired of the nasty names I've been calling him.

I've made no less than three trips since I began this little novel you're reading, so much water has gone under the bridge or many bridges over the water or something. Ask Gertrude.

I figure that to date I have personally delivered to our Allies, the Chinese, some 918,000 pounds of assorted canned beans, K rations, ammunition, gasoline, etc. Ho hum. Some days I shudder when I stop and think how this war might have ended were it not for me and Eisenhower. As I said before – ho hum. Being a modest chap I would naturally allow Eisenhower to take all the credit. Oh be quiet!

I just got in and I'm tired but it's been raining and nice and cool here so I'm my usual unhappy self. Honestly I don't believe I could be happy if I didn't have something to be unhappy about. Life would be too simple that way. Not that it's too complicated now but I'm getting a bit tired of all this work with no place to play. There'll be better days. I keep telling myself.

Honestly I'm so tired I gotta go now. Besides I have to get up at 4 A.M. tomorrow! @#%!@#%

It really doesn't make much difference whether my letters are long or short for I never say anything anyway. But if you'll promise to be a good little girl and write I'll make it longer next time. Say about 33 pages O.K.? Goodnight Now. Yours Always, Joe T.

By the way how have I been putting too little postage on my letters, are they too heavy or what? Do you mean it's going to cost me 12 cents to get these things delivered?

30 July

Hi Mary. For your information I just turned down an invitation to sit in a nice friendly little poker game so I could rush home to write this letter. You are a bad influence on my reputation, Miss Griffith, – you're curing too many of my vices. But don't rush out and buy a new halo for you've only just begun the good work.

It certainly is amazing the number of things a man can think of to waste his time. It appears to me I've been spending most of mine writing long pages of nothing to a woman I know. I'm in a rut. Sure is fun though so guess I'll stay there.

Tomorrow is the 38th anniversary of the Air Corp (ta da da-da, da-da!) so we're supposed to bust all records for Hump tonnage in a single day. According to the Director of Operations we're going to work so hard tomorrow we'll all have to take a week off for lack of anything left to haul. My back hurts already. But when doesn't it.

Believe it or not I saw a sure nuff American woman (white) at Kunming today. The first one in two weeks. I'm still shaking from the experience. I nearly broke my neck looking back. Of course she was merely a Red Cross girl but women is women and this is war. I'll bet even my Aunt Annie would be a glamour girl over here and she weighs around 275. I've got the damnedest bunch of relatives you've ever seen. All shapes and sizes. And all very boring. You must come to a family reunion with me some time.

It sez here in this nice long letter that came today (with the, quote "artistic," unquote, drawing enclosed) that I'm not to make any nasty cracks about the size of your feet. So I'll pretend to be the gentleman you know I'm not and refrain from comment. They aren't so big at that. I would love to be stepping on 'em right now to the strains of "Stardust" or "Smoke Gets in Your Eyes" or some appropriate thing like that.

Anyway I'm going to have a pair of mosquito boots made for you so I can kick you on the ankle under the table at post-war poker games without bruising you up too much. Thoughtful aren't I?

August 1945

1 August P.M.

Hi Mary. I'd better get this poor excuse for a letter in the mail before you decide I've deserted to the enemy and start thinking all those nasty things a lady shouldn't be thinking. At this point I might make a dirty crack but since I'm in a more or less happy frame of mind I'll be good and won't.

It's late and I've had approximately three hours sleep in the past two days and I wasn't on a binge. In fact I've actually been working for a change. Shuddup woman. I do [go] to work. Sometimes.

Anyway, I'd better go to bed now. I'll argue with you next time – say about five o'clock tomorrow afternoon when I get out of bed. Ho hum. I get sleepy just thinking about sleep. So Goodnight now – and write. Love and Kisses, Joe T.

4 August, 3 A.M.

Hi Woman. Most people would shake their heads and say, quote "tch tch," unquote, in that knowing way and go on thru life firmly convinced that I'm batty if they could see me stopping to write a letter at 3 o'clock in the morning. But I ain't like most people – fortunately. I'm just as firmly convinced that I'm right and the world is wrong – anyway I manage to have fun.

I've made four trips already this month so I haven't been exactly sitting around. We're always gone from twelve to fourteen hours. That's why I'm trying to squeeze in this letter now.

C-47 Douglas Skytrain

C-46 Curtiss Commando

If I went to bed the CQ would probably be around knocking at my door with the &!&&!@! little call sheet before I woke up.

I came in tired and dirty and disgusted tonight confidently expecting a letter at least 22 pages long after all this (sic) days. I look on my bed and under it and then in desperation turn the place upside down in search of that elusive letter and what do I find? Nothin'! – that's what! It's discouragin'!

Just after I send you some new furniture for that house of yours you up and inform me that you're going to be kicked out in the cruel, cold world again. Anyway those salt and pepper shakers you may or may not receive sometime this year I picked up in a place called Chengtu in China. one day. They are supposed to be made of some kind of stuff called Chengtu silver. Now as far as I know Chengtu Silver might be just another name for Pittsburg tin so don't blame me if they start turning a funny green color – I'm innocent.

4 August, 10:00 P.M.

Hi Topsy. Be quiet – I am too finishing this the same day I started! I didn't have to go tonite after all and it practically bwoke my wittle heart.

I finally got your 10 July letter today. I'm seriously considering writing the Postmaster General himself and telling him frankly just what a lousy organization he's running. Twenty-five days is too damn much! A July 19 letter came a week ago. I ask you – whereinhell have they been hiding this one?

I go to all the time and trouble to think up those vile things to think about you for not writing and all the time you're innocent and some bird brained postal officer is holding out on me. It's disgustin'!

Now lemme see – what have I been doing that would be interesting? Guess I'd better shuddup and change the subject before I start talking to myself again. I'm so bored with this cursed place I don't even appreciate the good liquor anymore. And lady, when I fall that low you can rest assured that I'm very, very bored. Ho hum.

The way I figure it I'll be shoving off for the Fatherland in November. ~~Hallulughyn~~ <– (can't spell that one) Praise Allah and all that there.

By the way have there been any illuminating news items in the Reed Gazette recently? I have just a faint (oh yeah?) idea (just my woman's intuition you know) there will be a featured editorial shortly. Subject: 'What a despisable (sic) bum Joe T. Pound is and how him and the gutter rat is first cousins.' I'll be practically ruined! So brace yourself and believe every word of it.

Had a letter from friend Charles S. Able this week. You know from the line of his conversation I don't believe the boy is very happy. He's drifting around the So. Pacific in some decrepit old tub called the U.S.S. Half Moon. According to him it should have been eclipsed sometime before the Spanish American War. Some guys are never happy. Just think how lucky he is down there surrounded by all those luscious hula girls you see in the movies. Tsk tsk. [For reasons never explained to me, Dad's friend who was named Charles S. Able always went by the name Bill Able. That's how I always knew him.]

I have sympathy not for anyone who hasn't served a tour of duty in the mystic (it sez here) land of India. Nothin' could be that bad.

Nope, I'm not very proud of the way I messed up my leave last fall. But I did not give you the run around as you tactfully put it. I was practically a prisoner what with my mama and (ahem) a few other people. Every time I started for a phone I'd get swatted down and have a piece of fried chicken or some of my mother's apple pie forced upon me.

Bob just came in from the club listing heavily to starboard. I think he must have scored a direct hit with a double rack of torpedo juice. I have him safe in drydock now so he'll be back at sea some time tomorrow afternoon. I have an idea he'll have a fire in the hold when he wakes up, tsk tsk.

I'd better haul down my mainsail and drop anchor in my bunk for ten or twelve hours. I won my water wings on the Wabash ferry at Merom so as one old salt to another — Goodnite Now. Luf, Joe T.

8 August

Hi You Big Beautiful Doll. Well, all right, if you're going to twist my arm I might consent to come back home to that old United States <—(spoken reluctantly - tsk tsk!) And just to show you how brave I am I'll even take you up on that dinner you're going to manufacture with your own two lily white hands. If I can live through our mess sergeant's nightmares of the past six months, lady, I can take anything up to broiled doorknobs and old rusty nails. Hmm. Chicken, did you say? With mashed potatoes and gravy?

In this here nice long letter that came today you have subtly insinuated that I'm fickle by suggesting that the picture I sent on its way to you followed some pretty girl off somewhere. How many times do I have to tell you that I'm completely impervious to beautiful women! My reputation speaks for itself. I swear I haven't given a second look to a female of the human species in two weeks The fact that there haven't been any females within miles for that period might have something to do with it.

I have to take a reassuring look at your picture every five minutes or so just to make sure I don't forget what pretty girls look like.

Had a letter from Gertrude today too. She says she's going to be generous when I come home and give me one night a week to go out with some other girl. Ain't she good to me? Of course she has no way of knowing that my mama only let's me out 'till eleven o'clock two nights a week.

Poor Junior is still in the hospital with his teeth wired together. I kept telling him he was too small to play baseball with the big boys but he took no heed of my kindly advice. So this Pirtle has a one track mind, eh? Just leave him to me (feel my muscle?) and I'll soon show him where he's rode to the end of the line and it's time to change cars.

Ex-marine eh? Hmmm. How big and tough is this guy? Perhaps we should go about this in a diplomatic way. Merely because I don't want to mess him up too much understand. Nine times I chased him around the block — and then he caught me.

You scare [the] hell out of me when you threaten to write me a sensible letter. Don't do it, please! I don't understand that kind of language. besides I would swear it was forgery!

My mama has gone to New York to see my sister who has gone there to see my brother-in-law who is a sailor. Complicated aren't it. I do hope my mama doesn't get lost in that there big city. My uncle Petey disappeared into a saloon in one of them cities two years ago and hasn't come out since.

The &!###xx lights just went out. So I had to fumble around and dig out our lantern we keep for just such emergencies. It's one A.M. anyway so I'd better say Goodnight — Correction, good morning. Anyway I just might find time to write again in a couple of days so don't think you're getting off so easy my dear — Love Always,
Joe T.

14 August

Hello Dear Mary. I'm way behind in my correspondence as a result of the excitement and confusion following a dozen or so false Armistices the past week.

Honestly when the war is actually over and someone rushes in to give with the glad tidings I'll probably just sit back quietly and fold my hands and close my eyes and grind my teeth and smirk, 'No foolin'?' for I'll be certain it's just another rumor.

At first Bob and I were inclined to tie one on every time the war ended merely as a tribute to such a momentous occasion but after the damn thing ended four times in as many days we ran out of liquor so perhaps we'll be drinking a toast to the Final Victory with a glass of lemonade or some such stuff as that. [Japan's surrender was in fact announced on 14 August, the date of this letter.]

Business has resumed as usual (Damit!) I was off on a trip all night last night and didn't bother to go to bed today so I'm not exactly feeling what you call sharp.

I'm just likely to cuddle up with my head on your nice soft shoulder and go to sleep any minute now — And while we're so chummy you can scratch my back — the part way up where i can't reach that always itches. Oh be quiet — I can dream can't I?

It was soooo nice of Leah to invite you up for the weekend. And what a rare coincidence that her two brothers (age 26 and 24) just happened to be home at the time. Hmmm. Are they Marines perchance? I hope you sunburn your nose and get sand in your ears. And may her two blasted brothers catch the measles! Hmpfff. (Long pause while I burn.) Now that I'm cremated to the proper shade of green let's get on with the story.

Lemme see. I got promoted this week. Or had you noticed. Which is another good reason the liquor situation is in a bad way. Nothing around but some paltry water.

Honestly, Mary, I'm practically asleep on my feet so I'll

say goodnite for now and if I have time write a dozen or so pages tomorrow. I have numerous topics to discuss with you – among which is howinhell Leah's two brothers decided to come home just when you decide to go up for a weekend. Mmmm.

And by the way – fear not whether I come thru Illinois or no on the way to Dayton. I figure about now me and an alley cat have about the same social standing in a certain Illinois community. Don't ask me how I know – It's just my woman's intuition, old girl.

I'll answer the rest of the questions in these two letters that came today tomorrow. (How's that for sentence construction? Egbert would hate me.) Goodnite. Love to You, Joe T.

23 August

Dear Mary. Today, 23 August 1945, I received a 1944 Christmas package and was so amused I decided to write and tell you about it since you always seem to be interested in these queer little quirks of Life etc.

Anyway that's the best excuse I can dig up offhand to give myself for writing you so here goes. . . .

I just received your letter of 13 August in which you wrongly accuse me of associating with one of those rare creatures known as women white. In the first place we have one (1) Ford staff car on the field for the use of all the officers for recreational purposes and there are so damn many playboys around here clamoring to use it you have to put your name on the list about three weeks in advance. Then you rush out and lay siege to some gorgeous little nurse until she finally breaks down and admits she ain't busy on this particular night for which you have the care reserved and agrees to have dinner at the Chinese restaurant and dance later. Then, quite elated, you run home and send out your nattiest uniform to the laundry and it comes back sans buttons. O.K., comes the evening of the big event. Along about sundown, while you're carefully shaving and showering and wondering if there's going to be a full moon tonite, the C.Q. strides in and announces (with a suitable leer on his face) that you're wanted for flight.

And so while you roar off into the black eared night with an unromantic load of canned milk and dehydrated potatoes Captain Paddlefoot comes along, takes car, takes girl, and has a lovely evening. [Paddlefoot was a derogatory term among aircrew for personnel who did not fly.]

It's happened time and again. The social life of a flying guy around here practically ain't. And the ground personnel weep and wail and tear their hair because we make 50% more money. Ho Hum. I feel sooo sorry for them.

We're going to keep flying just as if the war still was because we have (censored) thousand troops over in China who have a queer habit of getting hungry about three times a day. It seems the Navy scattered numerous mines through the waters of the South China Coast and until such time as they get around to picking 'em up so our ships can use the harbors we have to keep 'em fed over there. Just call me Grocery Packin' Papa.

Bob has finished his tour and [is] waiting [for] orders from Calcutta to go back to the States. He'll be leaving about the first of the month.

Junior came back from the hospital, stayed three days, and dashed off on leave to a rest camp up in the hills near Mt. Everest where, it is rumored, there are unlimited quantities of iced beer, big thick steaks, and inner spring mattresses three feet deep.

I recall that you once expressed a burning desire to see the Taj Mahal so I'm enclosing a snapshot taken from the air by a friend of mine.

Lemme see. What other troubles do I have to air. I got wet and have a sore throat and can hardly talk. The food has been positively horrible lately. I hate atabrine pills. Other than that I'm happy.

My mama and papa are visiting in New York and according to the letter I received today are having quite a gay time. I'm surprized Mom got Dad that far from home.

It's nearly midnite so I'd better get this in the mailbox and get in bed. I had to get up at ten o'clock this morning. Goodnight and write.Love, Joe T.

28 August

Dear Mary. I just today received this little volume written by you the night the war ended. It was very interesting. Both sentences! However, since it's you I'll overlook a few minor details such as the brevity of the thing and get busy saying hello right back at you.

Friend Bob has to rise and shine at 3 AM tomorrow so I'm

being nice to him for a change and writing in the basha of a friend of mine so he (Bob) can get some snoozin' done. He can never seem to understand how it is that I always burn the midnite oil writing letters to you after sleeping all day. He wants me to be rational and write my letters in the daytime just like ordinary people.

Of course I could go to a lot of time and trouble to explain to him that the correct psychological time to write to a pretty young lady is at midnight by candlelight but he would likely be too thick to appreciate these finer lines of reasoning so what's the use.[?]

Besides it's only five minutes past midnite and I'm not all that sleepy. I was out all night last night and slept all day today naturally. It's been raining and nice and cool for a change.

I wish you would hurry up and get settled down so I could stop worrying about you. It causes me no end of gray hair when I think of you possibly sleeping on a park bench somewhere.

Is Wayne home yet? Of course I don't suppose you really want to see him very badly but then it's always nice for his sake for him to get home.

I need approximately 20 more trips now to come home. I can get that in about two months or less. Maybe. The thing I like about the Army is the certainty of uncertainty. Do you follow me or shall I wait for you to catch up?

This is going to be sorta short for the longer I write the bluer I get. I'm a deep indigo hue right now. Now that the war's over I'm all for giving this damn place back to the Indians and taking a powder. I'd better get in bed and sleep it off before I start pulling my hair again. Goodnight Mary. Love Always, Joe T.

September 1945

4 September

Hello Mary. There was a miniature monsoon downpour here tonight and now everything smells clean and rain-washed and it's nice and cool and I'm in a relatively pleasant frame of mind so how can I go wrong in writing a letter to my pet brunette?

Yeah, I know, I've never yet gone right in anything I've ever done but be quiet, you, I might be the sensitive type.

I can't think of a single thing to gripe about – not even the food. So I can't even ask you to feel sorry for me tonight. We have a new Commanding Officer and shortly after he arrived on the field last week he discovered he was hungry. So he went up to the mess hall, naturally, took one look at the food, and screamed, naturally. So he called in the mess officer and built a big fire under him and he called in all the cooks and built little fires under them and you would be amazed at the improvement in the subsistence around here. Tsk, tsk! Nothing but Swiss Steak and Pork Chops and Fried Chicken these days. Don't you wish you were here? Well, so do I, but I don't think it would be a very nice place for a lady.

In my spare time I've been reading **Forever Amber**. Tch, tch! Such a naughty novel! But I hadda find out sometime. My mother never told me nothing! He said with that angelic smirk.

For your information, Miss Griffith, I made my excuses and bowed out of a very promising stag (naturally) party at the club tonight just so I could come over and answer this very nice and long letter I had from you today. Then, too, I'm going to have to fly pretty soon so had to stay in walking condition.

I dreamed of two blue orchids —- don't mind me, it's just this */##_??—**!! radio again. You know this typewriter does a remarkable job of swearing. Remember when Blue Orchids came out? I betcha we were having a swell time about then weren't we? And didn't know it. ["Blue Orchids" was a song written by Hoagy Carmichael and recorded separately in 1939 by bandleaders Tommy Dorsey and Glenn Miller.]

So Leah's OTHER BROTHER is home now eh? Tell me, how many brothers does she have for Pete's sake? Lissen, when you run in two on me at once I don't worry, much. But howinhell do you expect a little guy like me to contend with a whole housefull of Bukers?

The Hump isn't going to be closed for at least a couple of months so if someone comes a knocking at your door some sunny day soon don't scream and climb out of the fire escape . . . . it won't be me. At least not before November. Damit.

I trust by now you have received some of these little manuscripts I've been sending to Sullivan. This letter of yours that came today only took nine days. Which ain't so bad at all, considering. It used to take me a week to get letters in Texas. Ah Texas!

Personally, I think this is a very promising idea you have about going back home to work. Purely from a very selfish standpoint, of course. Then I wouldn't have to go so far when I was looking for a fight.

Speaking of fights . . . My Love Life has done struck a reef. Yep, I'm just a busted valentine. She jilted me. All of which goes to show you just how fickle women is.

It's getting awfully late and I really should do a little snoozin' before I'm called for flight. Be sweet and all that sort of stuff. G'night. Luf, Joe.

6 September Youknowwhere

Dear Mary. Y'know, I was just thinking that a year ago tonight I was sleeping in a hangar at Goose Bay, Labrador with a big snow bank just outside the door and the wind whistling around the corners and, ah, those invigorating icy drafts that sent shivering little fingers creeping under my six blankets! Next night we took off for Iceland in a blinding snowstorm. I shudder when I think how little

I appreciated that nice chilly weather. Curses. All the opportunities I've passed up in Life to appreciate things. Boy, how I'd love to go out and play in the snow again – and rub snow in your face. Remember those little lectures we used to get from Mr. Gilmore every year about throwing snowballs at innocent (?) little girls?

Junior is giving me a hard time these days. He was promoted to 1st Lt. last week and I haven't been able to do a thing with him since. And then, too, the Colonel is a bit disgruntled with friend Bob. It seems he (Bob) got slightly fragrant over at the club one night last week and heaved a table and a few other bits of furniture at some Captain he was currently mad at. So he had to make a formal visit at the Colonel's office at the Colonel's request next morning and the Old Man barred him from the club for thirty days and made him pay for the table and a dozen or so glasses and a few other fixtures that were damaged – not to mention the Captain's dented dignity. Tch tch! I don't know why the boy always insists on picking on the poor Captains. I think I'll move into a pup tent someplace before I fall under the evil influence of my boisterous roommates and stray off the paths of Rightousness (sic). Shuddup, brat. And since I think about it I Believe the word is spelled Righteousness.

What is Gertrude doing these days? I haven't gotten around to answering her last letter yet and she's being mean about the thing and won't write to me anymore.

This machine has a new ribbon and the whole thing looks messy but still it's more legible than my handwriting so quit gripin'. This is India you know.

I'm getting a bit tired sweating out going home and hardly have a nail left on either hand to chew. It's that bad. I finally had a letter from you yesterday and that helped some. Even if it was a short one. I'm only making a trip about every third or fourth day now that we're flying daylight only and the time in between drags on interminally (sic).

We have endless card games in the afternoons and I've read every book and magazine on the field. By the way, did I ever tell you that the name of this field is Mohanbari and it's located in the upper Assam Valley 'way up in the northeastern corner of India – sandwiched in between Tibet and Burma. The mighty muddy Bramaputra [Brahmaputra] River lies about two miles to the west and winds its lazy way eventually to the Bay of Bengal.

Incidentally this river is commonly referred to as the "Burma-Putrid." All this information won't be of any use to you, of course, but I'm telling [you] so you can make sure you never wander up in this corner of God's Malaria Mill. There is nothing here that would interest you, I assure you.

The country immediately around the field isn't as primitive as you might suspect. Most of the Valley is covered with British tea plantations. However we can fly twenty miles up the river and see herds of wild elephant and water buffalo. Of did I tell you this before?

There is a village called Dibrugarh a few miles down the highway. It's a very dirty, smelly, little spot and has a few shops that sell curios and ivory carvings and such stuff as that, two Chinese restaurants where the food is inspected by our medical officers. They are actually very clean and neat and I have no qualms about eating an occasional fried chicken there. My mother worries and stews and frets lest I'm not being properly nourished but she has nothing to worry about, really, but try and convince her about that. Some women are just naturally the stubbornest creatures . . . . (this last statement is made wiz ze raised eyebrow and a meaning glance in your direction.)

Take good care of Wayne. I don't imagine you'll have to tie him up to keep him from running off back overseas. Just ask a kid who has 366 days away from the old Estados Unides [Estados Unidos - Spanish, meaning United States]. He'll tell you what a wonderful place the rest of the world ain't. And just because your brother is home you don't have to drop everything and forget to write to me damit.

I think I've rambled on enough for one night. So be good and write and I'll be seein' Ya woman . . . . Goodnight. Luf etc.

15 September, 12:30 A.M.

Dear Mary. Yeah, I know, this is a helluva time of day to be writing letters but I don't get two repeat two letters from you every day and besides I'm not sleepy, so there, too, period.

I went out to fly tonight but my airplane didn't check out so after waiting four hours while they worked on it I got disgusted and cancelled. Which is why I'm out so late. And don't ask me if my mother knows I'm out, Cecelia.

I had a letter from the Sullivan County Information Center

today, namely my little chum, Mary Reed. All the local dirt was strictly third rate stuff this time. She must be slippin'. She was telling me about how cute were all the new babies in town and what color hair they had and what color eyes and all that, all of which interested me no end. Of course. I particularly like little girl babies – little girl babies about nineteen or twenty years old – I guess I'm just the paternal type that's all.

I finally got those boots for you today. I'll get 'em wrapped tomorrow and send them to your Sullivan address. They will probably hurt your feet but cheer up, Life's little trials and troubles merely serve to strengthen the character. (It says on page 37)

Tch! Tch! Miss Griffith! Such unladylike language in this letter of 4 September! I was practically shocked to pieces! With a little training I'm sure you would be capable of raising a few critical eyebrows around deah old Sullivan.

We had a directive from Wing Headquarters on the first of the month stating that as soon as we deliver 10,000 tons allotted cargo to China we could close the field and go home. So things are looking up. If everything goes smoothly we should be able to put that much over in about six weeks. Everyone is working like hell, even me, and morale has gone up about 70 percent.

You are causing me no end of confusion what with a new address and phone number every other week. I have two full pages (practically) of obsolete addresses and phone numbers of yours in my little black book. Such a women! You and Eleanor.

Mom told me that when she first arrived in Dayton, she and the other female office workers were housed in cottages and dormitories. She didn't care much for this arrangement and actively explored better alternatives. As co-workers came and went, the result was a series of moves.

Strangely enough I'm getting sleepy. I do get sleepy at the oddest times. Lemme see, it's only 1:07 A.M. Anyway I'd better run along. So Good Morning now. (to be continued) Luf and Stuff, Joe.

26 September

Dear Mary. I don't know just how far I'll get with this for I'm three down on the Flight Crew Schedule so as soon as three more airplanes get back from China — bang! Damit! We have 53 airplanes here and as soon as they come in they're gassed and reloaded and a fresh crew is called and they're on their way again. Just like the Pony Express. We're on the ground in China just long enough to get unloaded and get a cup of coffee.

I received a letter from you today — finally! I was sure you had been evicted again and had pawned your fountain pen to rent a park bench. But since such isn't the case and you've been an invalid and all that I'll just breathe a sigh of relief again and count the new gray hairs next time I shave.

I haven't written for several days — partly because I've been fairly busy and partly because the climate here just ain't conducive to strenuous activity — such as shoving a fountain pen around — in off duty hours. I have before been called what is known as lazy but I've never actually believed it myself until now and even now I attribute it to the damn weather. And I do mean damn. In addition to that, were it not a lady (it sez here) to whom I'm writing I would even go so far as to use a more forcible adjective in connection with the ______ weather.

I had a chance to stay here in India another year and fly for the airlines until they get their own organization set up and believe it or not I gave it some thought. (and shuddup. Of that I am too capable!) I sure as hell am not going to make $3,000 anywhere else next year. I'll be lucky if I make $30. However there are a few people Stateside I have a hankering to see before I get too old and decrepit to enjoy a good fight so t' hell wit' it.

By the way — do you think I might get a job sweeping the floors and emptying wastebaskets and pencil sharpeners in your office when I get my discharge in one hand and my tin cup and dark glasses in the other? Yeah, I know I would be more likely to get a job in the psychopathic dept.

I was going to write last night so I wandered over to the mess hall about midnight to get a cup of coffee first and got shooting the breeze with a guy I was up in France with. Seven cups of coffee later I look at my watch and it's 4 A.M. and naturally a little late to start a letter. So we'll blame this other guy for talking all night.

I'm going to have to finish this tomorrow. The C.Q. just cornered me with his wittle ol' call sheet. See ya.

30 September, Sunday Nite

Hi Brat – This is a particularly uninteresting evening. Bob is off somewhere getting drunk no doubt. And Junior is Officer of the Day and out cruising around the field in his jeep seeing that everything is orderly and peaceful and quiet and dull. I think he's not going to have to work very hard for nothing exciting ever happens around here. Except when a white women (sic) comes on the field. Then they ring the air raid siren and everyone rushes out to look – and whistle. Lest we forget you know. Oh me, I must write to Gertrude one day soon.

My mama and papa have gone to Virginia now to see my brother. They just got back from New York where they went to see my sister. I expect to wake up and find them on my front porch any day now. They never stay home any more. Home – what's that?

After long and careful consideration, Miss Griffith, I've decided that as a blonde you would look horrible. And besides I don't like 'em. I just naturally don't trust 'em. In fact I trust 'em even less than brunettes. There was a blonde in Texas who I was currently in love with while I was down there. I got a letter last week and she's getting married to some feather merchant. [Following World War II the term "feather merchant" came to mean anyone with a soft job who took life easy.] So don'tcha do it. Women is jus' naturally never satisfied. You're pretty enough with brown hair. And then there was a brickyard blonde who caused me a lot of trouble once. As I remember it she caused me a lot of trouble on several different occasions. Oh well – [Brickyard blonde refers to an American style blond ale.]

Starting tomorrow we won't be doing any more night flying. For which I was so thankful I faced the East and bowed to Allah seven times in succession. I have five trips to make and then home – maybe. Anything could happen and usually does in the ATC. Anyway I'll trust to luck and take you up on that football game. Provided you'll promise <u>not</u> to wear any of those hats you bought when you went shopping. Yeah, I like hats O.K. but not the kind women pick out.

I was just thinking that it's going to be awfully cold when I get home. I'll probably run out and catch pneumonia first thing. Ho Hum. Do you have any pretty nurses at the V.A.? Just curious. I'm currently on belligerent terms with friend Bob. He and a mutual friend named "Weak Eyes" Cushman came steaming in at <u>3 o'clock</u> this morning listing heavily to starboard and insisted that I get up and discuss the world situation with them. In the first place I don't know anything about the world situation and in the second place I wasn't eager to discuss anything I did know something about at three o'clock in the morning. So they discussed and I cussed.

It's getting sorta late so I better get showered and get in the hay. Write darn ya. Love Always, Joe T.

October 1945

17 October

Dear Mary. Yep, I'm still here – and growing more and more pessimistic about the whole thing. To me the most disgusting side of the situation is the fact that they didn't waste any time getting me over here but insist on fooling around indefinitely about getting me home. This Army is the only organization I've ever seen that could screw up consistently and do such a remarkably good job of it. Damn.

There was a time when I thought I'd be home for Christmas – then I set it up to Easter. Now I'm asking myself, Easter of which year? Ho hum. Thirteen more months in exile. It's appalling.

Bob is being disgustingly cheerful. He says the first two years overseas are always the hardest.

I'm taking a ship down to Agra tomorrow so I'll have something to keep me busy for a couple of days at least. It's about thirteen hundred miles to Agra so we'll be staying over night down there. This field is closed so we're giving all our airplanes to the Agra boys. I wanted to go down to see the Taj [Mahal] anyway.

I haven't had a letter from anyone for about three weeks so I'm very angry with the Postmaster General. Along with the rest of the eight balls who are running the Army. It's a mystery to me how we ever managed to win the war.

Anyway you keep writing for I might be here till hell freezes. Which reminds me, it's been much cooler this week. In fact a blanket at night isn't at all uncomfortable.

Ah India! So refreshing! So invigorating! So – I really shouldn't use so much profane language.

Is Wayne home yet? What I want to know is – how did he manage it? Maybe he could give me a few pointers.

I have to get up at 5:00 A.M. so I'd better get in the sack. Write – Love Always, **Joe.**

October 19, 1945

Dear Mary. It's about time I had a letter from someone – namely you. When I don't get a letter for three weeks I begin to wake up cross and irritable in the mornings and poor Junior leads a dog's life. This letter came in eight days which is remarkably good, considering.

I just got back from Agra a couple of hours ago. Those guys are really leading a rough life down there. Fifteen minutes after I landed I was sitting behind a steak this big.

We flew down in six hours and landed about two in the afternoon. Went to see the Taj Mahal which was more or less impressive. Later on we went over to the old stone castle where old Sha Jahan kept his harem of 1400 supposedly beautiful women. Tsk tsk! how times have changed! Now it's 1400 men to every woman. I came here about 400 years too late damit.

What do you mean, "if I consider squandering a bit of my time with you upon my return?" Whyinhell do you think I want to come home you bum? If you don't object I just might squeeze out a little free time to squander with you on Sundays. Also on Mondays, Tuesdays, Wednesdays, Thursdays, Fridays and Saturdays. All the rest of my time I'm going to spend being philosophical about the good old days in India.

Seriously though I don't know now when I'll be coming home. It might be next week or next year. But be patient please. I'm coming eventually so don't do any thing drastic.

Next day

Hi you – Friend Robert rushed me off to the second show last night so I didn't get finished with all the gossip. I had another letter from you today postmarked October 1. The one that came yesterday was postmarked October 11. What I want to know is – whereinhell have they been keeping this other one for the past ten days? Now they're even holding out my mail on me! I'm patiently waiting for that

20 page article you were going to dash off with which to torture me while your boss was in Chicago and you weren't busy. I hope the old boy catches the measles and has to stay three weeks. Not because I have anything against the guy personally but for selfish reasons of my own.

I think your idea to transfer to southern Indiana is the most intelligent (at this point I see an inviting chance to make a dirty crack but I'm being nice about the thing and keeping quiet) thing you've done yet. I've been wondering how I was going to manage it over to Dayton and back every night. Vincennes would be much better. Terre Haute would be better still. Things are looking up. On second thought, though, it might be a good idea for you to open up a branch of the V.A. in India. There may be lots of American troops requiring medical attention as a result of old age here while awaiting transportation back to the Promised Land. The war wasn't half as nerve wracking as the peace that followed. Ho hum, I can picture myself creaking down the gang plank with a cane and my long white beard blowing in the breeze.

I wrote a letter the night before I went to Agra but after searching around I discovered that I didn't have any stamps and Bob didn't have any stamps nor Junior and everyone else in the neighborhood is always asleep at the time of night I generally write my letters so I didn't get it mailed. So I'll put it in with this one.

I have to go down to the dispensary this afternoon to get a couple of arms full of shots which a[re] presumably necessary before redeployment to the U.S.A. That's one step in the right direction anyway. So I'd better rush down there and get it over with. I'll tell you tomorrow how sore my arm is. — Your ever lovin' Joe T.

October 25

Dear Mary. I've just been seeing three good friends of mine off for home. They went to Chabua in a truck to catch a plane to Karachi to catch a boat to the United States to catch a train for home. Rather Involved isn't it? I would even consider taking a pack mule if the way wasn't so wet in spots.

Of all the blasted inconceivable places to sit around and waste perfectly good time this is the last infernal place I would choose. The war wasn't half as tough as the peace that followed.

Maybe we'd better cancel that football game date until next season. I might possibly make it by then. Yeah, I know Mary . . . me too. Damit.

I had a letter from my mama today. She wants to know why I'm not coming home. Tsk tsk! Silly girl! If she were here I would probably have slugged her.

This is rather brief but I'd better go to bed and sleep it off. Goodnight you. Write. Love and stuff, Joe T.

Joe Thomas Pound and Mary Jane Griffith were married on 26 January 1946 in their hometown of Sullivan, Indiana.

Going Home

There is no extant letter dated after 25 October, but I know from other documents that Dad left India in November 1945, and the Hump airlift formally closed on 15 November 1945. I recall Dad saying that he had enough flying hours that he earned the option to fly back to the U.S. rather than take the usual (and much slower) return by ship, and that is what he did.

Years later - veteran pilot
My favorite of Dad's Air Force photos

Mom and Dad
Dad was recalled to active duty in 1968.
By this time he was Colonel Pound.

Epilogue
The Great Falls Road Not Taken

Following is a letter that Dad wrote as he was graduating from Indiana University after the war, Class of 1949. It is self-explanatory and raises the possibility of a wholly different life path that he never mentioned to me. I have no idea whether he ever sent a copy, and if so to whom.

> During the summer of 1946 I was assigned to East Base at Great Falls [Montana] while awaiting discharge from the Army. My wife and I met many fine people there, were favorably impressed with the city, and would like to make our home permanently in or near Great Falls.
>
> I expect to graduate from Indiana University this June with a major in accounting. Approximately one-third of my college credit hours have been earned in accounting courses. My scholastic average is 1.9 out of a possible 3.0 which places me in the upper 25% of a class of 1200. I have financed my education partly with the G.I. Bill of Rights, have been partly self-supporting, and have engaged in some extra-curricular activities.
>
> I am interested in securing a position as junior accountant with a responsible public accounting firm such as yours. I am 24 years old, of good appearance, and believe I have the personal traits necessary in public accounting.
>
> My business experience is somewhat limited due to my status as a student and 3 1/2 years in the Army Air Forces where I served as a transport pilot. I expect to take the C.P.A. examinations within the next two years after further study and practical experience. For additional information, I refer you to the enclosed personal data sheet.

I realize that you are at a distinct disadvantage in having no information to go on except the information I can provide by mail. However, if you have a position in your firm for which there is a good possibility I may be qualified, I am prepared to travel to Great Falls at my own risk and expense to talk with you. For personal references in Great Falls you may contact Mr. Morris Anderson, 808 Second Avenue North, or Mr. John B. Willits, 2301 Sixth Avenue North. You may also refer to Mr. Virgil Herring, Assistant Professor of Accounting, Indiana University, Bloomington, Indiana, and Mr. Austin S. Reed, Manager, Dome Gas Company, Sullivan, Indiana.

I will be happy to supply any other information you may require. Sincerely yours,

Joe Thomas Pound

In 1951 Dad was recalled to active duty in the US Air Force (USAF) and served in Korea as a combat B-26 pilot, Group Operations Officer, and Operations Analyst at Headquarters 5th Air Force. That period of active service also included stops at Fairchild AFB near Spokane, Washington, and at Fort Benning, Georgia. There are letters for that period too, but they will be another story.

In 1954 Mom and Dad settled in their hometown of Sullivan, Indiana and lived there until 1968, when Dad returned to USAF active duty and I left for college. Had I instead grown up in Great Falls, Montana, I probably would not have gone to college at DePauw University in Greencastle, Indiana, and — most importantly — I almost certainly would not have met my wife Sharon, who grew up near Sullivan in Robinson, Illinois. I have no idea what my Great Falls path might have been like, but I'm glad I took the one I did.

Mom and her mother
Bertha
1942

Dad's Army Air Force Cadet
portrait
1944

Acknowledgments

In writing this book as in my life in general, thanks goes first to my wife and life partner Sharon Pound, whose love and faith in me encouraged my best effort. Without her I'm not sure I would have risen to the challenge of writing the book.

Thanks also to High Tide Publications, Inc. founder and CEO Jeanne Johansen, who took a chance on a septuagenarian rookie author, and to Chief Editor Cindy Freeman, who made my book better.

Finally, thanks to the members of the Emerson Society of Williamsburg, who set an example of writerly excellence to help me see the path forward. I should especially thank Kathleen Decker and the late Ed Lull, who pointed me toward High Tide.

Any remaining flaws or errors in the book are my responsibility alone.

Index to Photographs

Bibliography

Alexander, Caroline. *Skies of Thunder —The Deadly World War II Mission Over the Roof of the World*, New York, Viking, 2024.

Binzer, Robert (as told to Rainy Horvath). *The Able Queen*, Scarsdale, NY, Perfect Storm Press, 2021.

Constein, Dr. Carl Frey. *The Hump —The First Airlift*, 1st Books, 2003.

Craven, Wesley Frank and Cate, James Lea, editors. *The Army Air Forces in World War II*, Chicago, University of Chicago Press, 1948-1958.

Daugherty, Leo J. III. *The Allied Resupply Effort in the China-Burma-India Theater During World War II*, Jefferson, North Carolina, McFarland & Company, Inc., 2008.

Dominick, Alexander S. *Flying the Hump —The War Diary of Peter H. Dominick*, Green Bay, Wisconsin, M&B Global Solutions, Inc., 2022.

La Vove, Arthur. *Hump Drivers —An American Pilot's Account of Flying Over the Himalayas During WWII*, Atglen, Pennsylvania, Schiffer Military, 2021.

Mitchell, William P. *From the Pilot Factory, 1942*, College Station, Texas, Texas A&M University Press, 2005.

Pike, Francis. *Hirohito's War: The Pacific War, 1941-1945*, London, Bloomsbury Publishing, 2016.

Spencer, Otha C. *Flying the Hump—Memories of an Air War*, College Station, Texas, Texas A&M University Press, 1992.

Thorne, Bliss K. *The Hump—The Great Himalayan Airlift of World War II*, Philadelphia and New York, J.B. Lippincott Company, 1965.

About the Author

Richard Eric Pound ("Eric") was born on 7 September 1950. He is the son of the late Joe Thomas Pound and Mary Jane Griffith Pound, both of Sullivan, Indiana. Like his parents, Eric is a graduate of Sullivan High School. He graduated from DePauw University in Greencastle, Indiana in 1972 with a B.A. in History. In 1973 he received an M.A. in History from the University of Illinois in Champaign-Urbana, Illinois. Eric and his wife Sharon, a talented artist, moved to Williamsburg, Virginia in 2017. They have three children and nine grandchildren. He speaks Swahili, French, and German.

Eric joined the Central Intelligence Agency in 1976 and retired from the CIA after nearly 34 years in 2009 with the rank of SIS-5 in the Senior Intelligence Service (roughly equivalent to "three star" military rank). He served 18 years in Europe and Africa in eight foreign tours of duty, five of them in command. He was Chief of Station four times and Chief of Base once. His headquarters assignments included global, continental and special operations responsibilities. In his post-CIA career Eric worked for Raytheon as Director of Program Development, Special Technologies and Operations, and as Vice President at Strategic Intelligence Group. He then enjoyed some time as a historian and mentor for junior officers and retired for the second time in 2017.

Eric was awarded the Distinguished Career Intelligence Medal, the Donovan Award, six Meritorious Unit Citations (five in command) and numerous Exceptional Performance Awards. The Greek government awarded him the Hellenic Coast Guard Medal of Honor and Merit. He was the only civilian and the only non-Greek ever to be awarded that medal.

Eric has traveled from Dar es Salaam to Zanzibar by open boat, visited the headwaters of the Nile River and, with Sharon, climbed Mt. Kilimanjaro. He is a past president of a chapter of the Central Intelligence Retirees Association. As of 2024 Eric serves as a National Park Service Volunteer at the Colonial National Historical Park, Historic Jamestowne. He is also a member of a local writers' group in Williamsburg, the *Emerson Society*.

Made in the USA
Monee, IL
12 September 2026

62976626R00076